Feeding of Organic Farm Animals: Pigs, Poultry, Cattle, Sheep and Goats

Feeding of Organic Farm Animals: Pigs, Poultry, Cattle, Sheep and Goats

Edited by:

Vijay D. Kele and
Dr. Prafullakumar V.Patil

Delve
PUBLISHING
www.delvepublishing.com

Feeding of Organic Farm Animals: Pigs, Poultry, Cattle, Sheep and Goats

Vijay D. Kele and Dr. Prafullakumar V.Patil

Delve Publishing
224 Shoreacres Road
Burlington, ON L7L 2H2
Canada
www.delvepublishing.com

Email: orders@arclereducation.com

ISBN: 978-1-77469-091-8 (Hardcover)

ABOUT THE EDITORS

Dr. Vijay Kele is presently serving as Associate Professor, Department of Dairy Technology and Food Technology, PIT, Parul University, Vadodara, Gujarat. He obtained his B. Tech (Dairy Technology) from MAFSU, Nagpur, M. Sc. (Dairy Science), MBA (Food & Agribusiness Management) & secured 04 gold medals from University of Mysore, Mysore and Doctorate in Dairy Extension Management from Nagpur. He having 11 years of Teaching and 03 years of industrial experience. He has published 04 books and 18 research paper in National and international journals of reputed, popular article 87, extension article 72 however 14 radio talk and 02 television live talk on Doordarshan. He has participated many National, State Seminar and symposiums. He has written 22 practical manuals for under graduate course for the benefit of students besides Lectures in farmer traning programme, organized 133 farmers scientist meets, organized 14 exhibitions. He has organized 73 webinars, 07 STTPs (one week), 12 Workshop and 01 FDP. As per research concern received grants for 04 projects from ICAR, Gujcost, SSIP (Govt. of Gujarat) However 02 Patents in his credits.

Dr.Prafullakumar V.Patil, presently working as a Farm Manager at Veterinary College, Udgir. He did his Master degree in Animal Nutrition in 2006 from MAFSU, Nagpur. He has qualified ICAR, NET examination in 2010. Now, pursuing PhD in Animal Nutrition at the College of Veterinary Science and Animal Husbandry, Anjora, Durg, Chhattisgarh. He is actively engaged in research, extension and clinical activities for the last 15 years. Also engaged in teaching from the last five years. He has published 37 research papers and 7 review papers in national and international journals along with 242 popular articles. He has received the best poster presentation award in the national symposium at PGIVAS, Akola (Maharashtra) in 2019. He has received a number of honours/awards for best extension work in the veterinary field. He has organized more than 21 trainings for field veterinarians; farmers etc. He has organized and participated in more than 150 farmers meet along with 45 animal health camps. He has authored two textbooks viz.'Handbook of Animal Nutrition' and 'Milk production management' in English and two books in Marathi. He has participated in more than 80 training programmes as a resource person and also conducted more than 130 demonstrations for field veterinarians, farmers and unemployed youths. He has also prepared an album of different feed ingredients and fodders. He has also acted as editor for 6 booklets on 'Profitable goat farming', 'Profitable dairy farming', 'Scientific poultry farming', 'Backyard poultry farming', 'Fodder production and management', Nutritional management of animals during scarcity period'.

TABLE OF CONTENTS

LIST OF FIGURES

ABSTRACT

Organic animal husbandry is a valuable strategy for overcoming the setbacks of the agricultural industry (sustainability, food security and safety), while at the same time conforming to consumer preferences (animal wellbeing, health, environmental safety, etc.). The volume introduces organic animal husbandry as an interesting strategy to solve the developmental strategy of rural areas and the decline in farm profits. The volume mentions that animal husbandry is vital to the sustainability of the rural economy and many ecosystems. However, it also has a high environmental impact. Due to the growing demand for animal products, new livestock production systems need to be designed to balance food security and sustainability. Some chapters covered in this volume include chapter 1: an introduction to organic livestock, chapter 2: factors influencing organic livestock, and chapter 3: organic livestock feed and their benefits among others.

Organic livestock may be a valuable strategy to realize agricultural sustainability goals. Nevertheless, there is a lack of research on integrating existing knowledge, especially knowledge of organic livestock, into the main aspects (its externalities and challenges) implicit in its practice. The volume aims to fill this knowledge gap and provide strategies and insights that will help stakeholders and policymakers improve the sustainability of organic food and the entire food system.

PREFACE

In the past ten years, the production of organic livestock has grown rapidly in all countries. The number of organic livestock producers is still growing, although not at the rate of the late 1990s, and the increase in organic agricultural subsidies is likely to maintain this steady growth in organic livestock production in the next few years.

This volume indicates that organic farming is based on a closed agricultural system and minimal use of non-renewable energy sources (such as artificial fertilizers). Livestock, especially ruminants, play an important role in maintaining the fertility of grass/clover. The basic principles of organic livestock production can be summarized as follows: land activities (self-produced feed; manure returns to the same land where feed is produced); good animal health and welfare (outdoor, free-range, maximum use of pasture, etc.); optimization instead of maximizing production (breeding for traits other than yield or growth rate, feeding of specific types of feed, etc.). Compared with conventional systems, population density and yield levels are lower (the population density of dairy farms is reduced by an average of 80%; beef, and sheep farms are reduced by 40-80%).

The volume covers the following areas of animal husbandry production and animal husbandry: the origin of animal husbandry, animal husbandry conversion, animal health and welfare, housing and animal husbandry, feeding and livestock records. These topics are covered in 10 chapters which include; Chapter 1: an introduction to organic livestock, chapter 2: factors influencing organic livestock, chapter 3: organic livestock feed and their benefits, chapter 4: limitations of organic animal production, chapter 5: organic pigs, chapter 6: organic poultry, chapter 7: organic beef, chapter 8: organic sheep, chapter 9: organic goat and chapter 10: markets of organic livestock.

If animals are to be slaughtered to produce organic meat, they must be born in organically managed dams. The volume suggests that cows need a conversion period of at least 36 weeks. If the calves are to be sold organically, they should be managed organically for 12 weeks before they are delivered. During the conversion duration, the feed must be done with organically produced nourishment and managed following organic standards. As long as the livestock conversion plan is developed and implemented from the start of the conversion period, land and livestock conversion can be carried out at the same time.

In the treatment of clinical diseases or when there is a known disease risk, veterinary drug input such as antibiotics, antiparasitic drugs and vaccines are acceptable in organic animal farming. When preventive or preventive input is used on a herd/flock, the veterinarian who prescribes is usually required to identify specific risk factors and explain why it cannot be handled through livestock/system changes (e.g. double fencing and prohibition in the herd/ preventive use of antibiotics on the flock).

Many countries adopt national or "private" standards that do not have legal status. The International Federation of Organic Agriculture Movements (IFOAM) has developed a certification program for private and public certification bodies intending to bring standards (especially the production of export commodities) to uniform standards. Currently, the practice of organic agriculture is a farm assurance standard defined by EU and US laws. The EC Regulation 1804/99 on organic livestock production came into effect in August 2000.

Chapter 1

An Introduction to Organic Livestock

CONTENTS

1.1 A BRIEF OVERVIEW

Livestock is recognized as animals that are domesticated and are raised in an agricultural setting. In most instances, these animals play a vital role in offering household labour especially tilling the land, these are the bulls. Other roles played by these animals include the provision of commodities such as meat, milk, eggs, fur, leather and wool mainly from the sheep. The word livestock is used to mean differently in different settings. In some settings, for instance, the term is used to mean those animals that are reared for the sole purpose of consuming them while in some instances; the term is used to define farmed ruminants such as goats, sheep and cattle. In some areas such as the United States, horses are also classified as cattle.

Figure 1.1: Livestock and poultry. Different organizations have different ways of defining livestock.

Source: V Drovers

According to USDA, animals that are categorized as livestock include pork, veal, beef lamb and animals with red meat. According to this organization, fish and poultry are not considered livestock. From the time when humanity transitioned from hunter-gathering to farming, animal husbandry has been a very important component that is being practicsed in most cultures. Animal husbandry involves breeding, maintenance and the slaughter of animals. These practices vary in different settings and cultures but are rather seen to play a vital role in the economic and cultural development of a community (Wilhelmsson *et al.*, 2019). The mode of rearing animals can be either intensively or extensively. The extensive system is also referred to as the organic livestock farming system. This system involves animals roaming around in an agricultural land of their own free will. The roaming is normally under the supervision of a herdsman manly to protect the animals from being captured by predators. In the western United States, for instance, ranching normally involves a large number of cattle that are seen roaming around in either public or private lands. This normally happens in areas where there are large pieces of land and the amounts of rainfall are minimal. In areas where there is snow, rearing of cattle proves to be very difficult as they can die during the cold winter. However, in instances where there is the rearing of animals in areas that experience winter, the animals are brought out during the spring and stay inside during winter and the stored pasture is provided together with other supplements. In rural locations and the African communities, extensive mode of rearing animals is very common. Take for instance chicken in the African setting. The chicken may not be fed but left to roam in the fields and surprisingly it will be able to produce either one or two eggs in a week.

In developed countries, however, the majority of the animals are intensively kept. It is common especially for the dairy cattle that are normally zero-grazed and the forage is brought to their feeding stations. In between the intensive and the extensive forms of livestock keeping is the semi-intensive. This normally happens in a family in which the livestock are allowed to freely roam outside for a period in which there is enough grass to feed them. However, when the grass is depleted, the animals are taken back to their confinements are forage is brought to them. During the period in which the animals are in a confinement, the grass is allowed to grow to a height that can be able to feed the animals (Ankamah-Yeboah *et al.*, 2019).

When the intensive form of livestock rearing is adopted, the animals are always at higher risks of contracting diseases. Furthermore, after a single cow has contracted a disease, there will be a likelihood that it will spread the disease to other livestock within the confinement area. The case with contracting diseases is very common since the rearing animals in a confined space may prove to be unhygienic. Maintaining confinement can be very difficult and is time-consuming. In case a farmer has chosen this form of livestock rearing, then they should be ready to maintain high levels of hygiene in the confinement spaces.

1.2 ORGANIC LIVESTOCK MANAGEMENT

Organic agriculture, in general, is considered as a good production system with a number of benefits including the promotion and the enhancement of the ecosystem, health, biodiversity, biological cycles and improvement of the soil biological activity. The emphasis is normally on the management practices that utilize inputs, which are preferably obtained off-farm. Farmers are required to take into account the fact that the regional conditions require systems that are locally adopted. If possible, the farmers are advised to use mechanical, agronomic and biological methods instead of using synthetic methods in order to accomplish what they need to do.

Figure 1.2: Organic dairy farming where dairy cattle are allowed to roam the fields to get pasture on their own.

Source: eOrganic

Organic animal husbandry involves the production of livestock that utilizes organic and biodegradable inputs that are obtained from the ecosystem of animal nutrition, animal housing and breeding and animal health. This mode of livestock management deliberately avoids the application of synthetic inputs while rearing livestock. Such synthetic inputs include genetically engineered breeding inputs, drugs and feed additives. In countries where the farmers are constrained in terms of the resources available to them, they tend to use a limited number of external inputs such as allopathic medicines. However, their mode of feeding animals involves animals freely grazing. To a larger extent, such farmers are considered to be much closer to organic farming, which normally happens, by default.

In most instances, these farmers fall under the category of those that do not have efficient knowledge regarding organic farming and thus limiting them from exploiting the various benefits of organic farming. Even with this situation, these developing countries are becoming the most important suppliers of organic foods. This is so because the majority of the producers in the developing countries practice organic farming especially when the farm size is small and the areas experience substantial amounts of rainfall.

In the current society, there has been increased consumption of organic products, especially in the developed countries. These individuals are willing to pay a premium price to acquire these products (Åkerfeldt *et al.*, 2020). As such, the organic products market has become a great niche for exploitation by individuals from developing countries. Farmers in developing countries are therefore encouraged to rear their livestock in an organic manner, which helps the soil to gain nutrients. When products are planted in such a land, they will not need to apply any form of fertilizer that makes their produce potentially organic. With available markets, farmers will be able to increase their income from the farm which improves the gross national product or rather the net income in the country.

Figure 1.3: Advances on organic crop production has been done with little on organic livestock production.

Source: Refix Commodities

When compared to conventional livestock, organic livestock is managed differently. With organic livestock, there are normally restrictions on the use of antibiotics and hormonal improvements, which then benefit the environment and people. The emphasis for organic livestock is on the use of pastures. Those animals that are raised in organic terms tend to be better shape than those raised in a conventional setting. Furthermore, organic livestock are normally good or rather better for consumption in terms of the health of the consumers.

In most parts of the world, advances in organic crop production have been done with little development of organic livestock operations. The reason behind this fact is that there is increased demand for organic crops unlike for organic livestock prospects, which are limited owing to the various quality controls otherwise known as the trade barriers. Another reason is that most of the developed countries are self-sufficient in terms of organic livestock products. Furthermore, organic livestock products are traded between the developed countries leaving the other minor countries with no opportunities for trading. The quality standards for animal products have some form of disparities depending on a particular country making it hard to trade for such stated products.

Unlike the conventional production systems, organic systems of production are governed by certain standards that should strictly be followed by the producers of the products. To show that the producers have adhered to these standards, they need to get certification that guarantees the quality of the products and the production process. Since organic products are identified through their labelling and identification, they are considered as the consumer or market-driven products. Consumers are normally conscious about the production of the food that they ingest in their bodies. As such, the producer needs to manage their organic livestock production systems in a manner that meets the various stated standards to acquire a market for their products. Any farmer who has the will to follow the various stated principles can practice organic livestock farming and the guidelines stipulated on the various ways to manage organic livestock systems (Alonso *et al.*, 2020). This applies to all farmers, whether they are producing the product for home consumption or marketing purposes. However, for the farmer who is practising this agricultural system to sell the products in the market, they need to get a certification to show that they are following the guidelines. The certification is normally provided by an accredited agency and the various steps to be followed to acquire certification include:

- Registration of the organic livestock producers and the various processing industries
- Provision of the basic information about the livestock reared on the farm

- Inspection and the verification of the field and the processing unit in which the organic livestock are processed
- Inspection of the methods of production and practices used by an inspector that has been appointed by the certifying agency

When transitioning to organic livestock production, the producer is given 1-3 years during which synthetic inputs are not to be used on the livestock before the stated producer has his products identified as organic. The most important goal of organic animal husbandry is ensuring that the organic livestock have good health and welfare.

1.3 KEY CONSIDERATIONS IN ORGANIC LIVESTOCK PRODUCTION

With the changes in consumption patterns in the world, there will be increased demand for organic livestock products. In order to meet the various principles and guidelines, there are some key considerations of organic livestock husbandry that the producers should have in mind.

Origin of the livestock- the livestock and the products from the livestock labelled as organic require to be raised in continuous organic management.

Livestock feed-rations that are produced under the organic management development should constitute agricultural products that have been organically produced. Furthermore, handling should be done organically. For feed additives and supplements, the farmer may use certain synthetic non-synthetic substances. Take for instance dairy cattle under the age of 9 months; twenty percent of the feeds can come from non-organic substances.

Living conditions- an organic livestock farmer is required to create and maintain the living space of the livestock to ensure that there is proper hygiene and that they can be able to accommodate the natural environment of the livestock. In such living conditions, the livestock should have the ability to access fresh air, shelter, access outdoors, shade, access to pasture for ruminants and direct sunlight for particular species of animals.

Waste management- farmers that are involved in organic livestock farming have an obligation to manage manure in order to prevent the contamination of the soil, crops and water. The manure should be used in a manner that optimizes the recycling of nutrients.

Health care- farmers practising organic livestock farming have an obligation to practice preventive health care. First, they need to select the appropriate species of livestock, provide feed that is deemed adequate, create an environment that minimizes diseases stress and attacks from parasites, make sure that they administer vaccines and veterinary biologics and follow the various animal husbandry practices that promote the wellbeing of the livestock (Vogl *et al.*, 2016).

Record keeping- there are a number of reasons why producers of organic livestock are required to maintain records for their systems. Financial management of the enterprise is one reason why records need to be kept. Furthermore, records are also important for verifying the organic status of the animals, production, handling and harvesting practices.

1.4 IMPORTANCE OF PASTURE

Figure 1.4: Hay and forage are good organic feeds that can be very profitable to the producers.

Source: ICS The Partner of your Farming Projects Africa Middle East

Good pastures are very profitable whether the producer is a commercial one or a part-timer. They provide an economic source for building soil tilt and fertility, reduce labour requirements, reduce erosion, provide livestock feeds and reduce the invasion of poisonous and noxious weeds. When one needs to use a new land for the purpose of forage productivity, the farmer is asked the first year into careful consideration. If they manage the land appropriately and use the fertilizer wisely, there will be a high chance that they will learn the capacity of the land with regard to forage production. In the case where a farmer establishes a pasture from the grass root, the process may sound a bit thorough. However, if not properly managed, the pasture may easily revert to the previous plant composition after a few years. This may prove to be a waste of money, time and labour. If the perennial pastures are properly managed and established, there is a higher chance that there will be an effective product in the next 5-10 years. A producer of forage needs to have the ability to establish a forage stand in the situation where the pasture or forage crop is in high demand but is unavailable.

To start from scratch means that the farmer will have to recognize land that was planted initially in a monoculture crop. Otherwise, the producer may opt for an area, which was initially wooded but has been cleared up. The business of establishing a forage plot for the sole purposes of pasture or harvesting in a mechanical manner is described as a very risky business. In order to remove what is growing in particular land for planting forage should be done with careful thinking and preparation.

In general, establishing pasture means the production of forage of high quality that is meant to meet the intended livestock. It is with no doubt that a producer can carefully plan what he regards as best for his operation mainly because they are in control. Years of high-quality feeds can be achieved if the forage production operation is well planned and carried out in a manner that is efficient for the forage. The producer needs to understand that they need to do proper planning, preferably a year before entering into the particular enterprise. There are also specific steps of operation that need to be carefully followed. The entire process may however be sabotaged if the weather conditions are very harsh. Since there will be a time when there will be no plants holding on to the soil in the particular land, there might be cases of soil erosion, which may negatively affect the land (Bessei, 2018).

When pasture is used to feed the livestock, one may discover that there will be fewer visits and calling of the veterinarian. Bills will also be reduced and further, the health of both animals and land will be improved. The animals will also be able to provide the setting with fertilizer since the farmer can use their dung as manure. Since the livestock will be feeding on the land that is being tilled, then there will be a reduction in weeds. The environment in which the animals access their diet is normally a natural setting. Furthermore, the animals will be able to access a variety of diets that will improve their health and wellbeing.

1.5 HERITAGE BREEDS

These are traditional breeds that were originally raised by our forefathers. Researchers claim that they are breeds of a bygone era, that is, they existed before the process of the industrial revolution. They existed before industrial agriculture became a practice in the world. The reason why these breeds became well adapted to the local environment and cultural conditions was that they were carefully selected and bred over time in order to develop such traits. These animals thrived in conditions that are very different from the animals found in modern agriculture. These traditional historic breeds had the capacity to be self-sufficient and their survival rates were very high. Furthermore, they had the ability to produce manure that would sustain the croplands in the region, lived for longer periods, had maternal instincts and had the ability to mate naturally. They were also very resistant to diseases and parasites. There was a time when the heritage breeds were seen to roam freely in the pastures of America's landscape. However, in the current society, these breeds are on the verge of extinction because of the various changes done to modern agriculture, which makes these breeds fall out of favour with the environment (Buller *et al.*, 2018). There is a wealth of resources within the heritage breeds that make them considered important for the future of the society and that of the food systems.

Figure 1.5: Heritage breeds are the best types of animals one can ever own as they are resilient to most factors.

Source: Off The Grid News

In order to codify the term heritage in a marketplace, the Livestock Conservancy is working towards coming up with a definition for the same. As far as agriculture is concerned, heritage for the turkeys, chickens, swine and cattle have been defined. In order to understand the term heritage, in simple terms, one would find these animals on their grandparents' farm. To define the pure breeds of chicken and cattle with histories that are deep within the context of the United States, heritage represents the umbrella term. These are animals that were bred to meet the local conditions. The breeding was done over time in order to get the best results for the animals. Since the selection and breedingare done over time, they tend to have better resistance to diseases, are well adapted to environments and are considered to thrive well in settings that are based on pasture. In the current society, there are breeds of livestock that have been selected for large-scale agriculture for intensive production purposes and have characteristics such as being feed efficient and continuous production of products. Heritage breeds on the other hand have characteristics that are well rounded.

The reason why the livestock conservancy is working hard to define heritage on livestock and chicken is that it may become a term that consumers can rely on when buying products in the food and agricultural marketplaces. By defining heritage, it helps create a market niche that will help in the conservation of the traditional breeds in modern society.

In the current society, most animals have been bred to survive in harsh or rather cramped conditions. This is normally for the case in which a farmer has a large livestock operation. In this case, the animals tend to grow quickly and the quantities of meat are normally higher compared to livestock in their farms. In most cases, organic farmers who feed their animals with pasture prefer raising heritage breeds (Broom *et al.*, 2013). The animals tend to be very efficient since they tend to be hardier and are suited in a way that they can be able to survive life outdoors. In Rodale institute, for instance, the organization chooses to raise heritage hogs for the sole purpose of research. The reason for this is that, when comparing these breeds with the modern-day breeds, they are considered easy to rare since they are predisposed to forage. Furthermore, they require little attention and may prove to be healthy most of the time. Changes in climate also have no severe effect on the heritage breeds.

1.6 WHY ORGANIC FARMING MATTERS

Considering that the other type of farming is factory farming, we can say that organic farming is the best fit for our environment. The environment is always changing and when it is not taken care of, we might end up on the losing side. Take factory farming for instance. Factory farming has been reported to be one of the major contributors of the climate change in the world. In this case, most of the animals are raised in a confined area, which is forcing individuals to clear out areas covered by forests. By clearing out the forest spaces, individuals are trying to make room for the buildings in which the animals will be living.

Figure 1.6: Intensive livestock farming has a negative effect on humans, animals and the planet.

Source: sentient media

With this case scenario, the area that is covered by forests will be reduced which will lead to soil erosion and drought. In a way, forests are some of the natural features that contribute to rainfall. With forests gone, the world will be faced or will rather be left in a situation of drought. When animals are confined in a single area, the pasture in the particular space will be depleted leaving the land vulnerable to wind and water erosion. That is also not the worst-case scenario. With the confinement, the manure will be collected in pits or rather assembled in a particular area, which leads to the release of nitrous oxide, potent greenhouse gases and methane. These gases are not at all good for the environment; they are rather responsible for the destruction of the ozone layer, which exposes the world to amounts of heat that are harmful to human beings.

In the current society, most individuals are shifting towards factory farming or what is commonly known as intensive animal farming. Take for instance in the United States, the majority of livestock; approximately 99% of the animals are reared in this manner. The positive impact of this type of farming is that there have been increased yields for commercial purposes. As already established, the negative impacts of this type of farming surpass the positive impacts (Brito *et al.*, 2020). Consequently, the negative impacts are very harsh as they considerably affect the environment, the welfare of the animals and the welfare of the public. Because of the above-stated negative impacts experienced because of practising factory farming, researchers claim that in the years to come; there will be a large decline in livestock. Precisely, the claims are that by the year 2030, animals such as cattle will decline and by the year 2100, the will be an end to animal husbandry in general. The animals that are raised in a factory farm are fed grains in order to gain weight in a speedy manner. Grains are very dry and thus these animals take in a lot of water. This is unlike the animals that are pasture-fed. Pasture normally contains some amounts of moisture and therefore the animals do not drink a lot of water but rather acquire it from the pasture. Furthermore, cows and other ruminants are not built in a manner in which they can digest grains. As such, these animals are normally faced with a number of complications. Farmers are then forced to increase their use of antibiotics, which leads to animals facing a threat of "superbug" outbreaks. In order to meet the demand for grains, other farmers practice monoculture farming of corn seeds and soy seeds. This leads to the rapid degradation of the land as the nutrients in the soil are slowly lost. By planting one type of crop on a particular farm, the nutrients in that soil will be depleted. In order to restore the amounts of nutrients in the soil, farmers are then forced to use fertilizer, which in a way is very harmful to the soil and largely harmful to the environment.

Figure 1.7: Animals raised in a factory farm are fed grains, which may be harmful to their health.

Source: GRIT

Since organic livestock farming is based on feeding animals with pasture, it promotes the humane treatment of the animals, soils remain healthy and most important, and it promotes biodiversity. Animals pick their feed from the agricultural land and in the process may come across some plants that are considered to be very beneficial to their bodies. Furthermore, animal waste is used as a form of manure. Manure that is from the animals is normally non-hazardous to the environment and is very great in terms of improving the fertility of the soil. Soil with animal manure that is used in planting improves the yield of the products that have been planted. Considering all other forms of farming practices, livestock operations that are pasture-based are considered to have the most potential to release carbon (Cambra-López *et al.*, 2010).

1.7 WHY CHOOSE ORGANIC LIVESTOCK

Organic livestock has been embraced globally by local farmers and other farming institutions as studies depict that animals fed with organically grown feed have better growth and reproduction growth rates than conventionally grown feed. Generally, it is assumed that animals have a better immune response towards prevailing pathogens under organic conditions than under conventional conditions although this has yet to be scientifically proven. Many scientists have argued that important aspects of organic farming, like genetic background, nutrition, environmental exposure, weaning age and stress may affect the immune response of the animals and should preferably be analyzed separately in ensuring the selection of certain immune response traits relevant to organic farming feeding investigations on immunological defensce in animals.

Declining and stagnating losses in organic livestock are rarely experienced due to some fed with local herbs that are seen as providers of chemical compounds that reduce the incidence of certain diseases particularly those caused by internal parasites in livestock. Moreover, forages contain other compounds that help to modulate rumen microbiota and increase feed efficiency. This is the case for alfalfa plants, high in malic acid content, which help to reduce the incidence of acidosis, and increases diet digestibility and milk production. Moreover, some legumes contain useful compounds for parasite control. Rumen microbiomes are complex and interact with dietary compounds. Therefore, the use of products that in a 'natural' way allows to both modulate rumen microbiology and normal rumen functioning is required for the high production levels to achieve food security.

Feedex, an organic livestock feed manufacturer in the Midwest that steam flakes our grains advice that regularly feeding animals with the loaded nutritious organic feed eventually would result in healthier animals. This is because of the tremendous nutrients such as minerals and vitamins found in organic feeds. Additionally, organic livestock feed helps with digestion, strong bones, muscle development, skin and coat health, energy levels and lactation. Organic livestock farming has assisted farmers in saving valuable resources. Organic livestock has higher starch consistency, which requires farmers to feed their animals less. The animals gain more weight per pound of food while improving their efficiencies to about 20 percent.

Figure 1.8: Animal nutrition and feed rations in organic livestock, forages contain other compounds that help to modulate rumen microbiota and increase feed efficiency.

From an economic point of view, studies have reported that feeds grown on-farm for livestock consumption are more environmentally favourable than imported feeds, partly due to their lower energy requirements for transport and lower greenhouse gas emissions. Due to the cost of protein, it may also be economically beneficial for farmers to include more homegrown proteins in dairy cow diets than the commercial supplements. Due to this, different authors have considered that although the integration of crops and livestock is essential for the economic sustainability of pasture-based ruminant systems, the local context makes it necessary to analyze the financial effects of such integration and its optimization under the specific farms' conditions for better estimations (Castellini *et al.*, 2012). Observations have been made that low input grazing-based systems offer higher prices to livestock producers for their livestock and by-products. However, the ongoing regulations on organic production in the 100 percent organic feed rule will increase the marginal feed input cost, making it unprofitable to keep the same number of cows, which forces farms to make profound management and structural changes. The organic systems that are based more on pastures would definitely suffer less.

Moreover, self-sufficiency and the reduction of feed cost have been identified as essential for pasture-based ruminants' farms in organic livestock farming. In cases in which milk production is associated with a lower use of external feed, the farms' economic sustainability may be positive if the ratio of input/output is beneficial and the farmers still obtain enough income. Moreover, the lower use of feed would be in line with organic principles. This higher self-sufficiency was also identified as positive by researchers, as the organic system offers opportunities for reducing nitrogen surpluses. However, soil management should be improved, as they detected the risk of phosphorus and K potassium deficits and the need for additional fertilizer. Once more, the importance of adequate integration of livestock and crops, as well as correct grazing systems, is highlighted. There has been a further collaboration between arable farming and organic livestock production for manure exchange. Several studies have stressed the need to reduce inputs in organic livestock farming and maximize profits by producing high-quality forages and increasing the dry matter on the pasture.

Minerals as nutrients have been studied extensively in organic dairy farming, due to their influence on animal health through immunity. Trace minerals, especially zinc, are essential to maintain the normal function of the immune system in ruminants. Copper has been related to an increased resistance to mastitis which is one of the most relevant problems in dairy animals, particularly that caused by *Escherichia coli*. Reviews on the importance of trace mineral nutrition in organic livestock farming found that the use of chemical forms of minerals such as organic minerals or chelates that are effective in production performance and health must be considered. Trace minerals are chemically bound to a chelating agent or ligand, usually a mixture of amino acids or small peptides. This makes them more bio-available and bioactive and provides the animal with a metabolic advantage that often results in improved performance. Organic livestock feed provides animals with necessary nutrients without all the harmful chemicals, hormones and fillers commonly found in non-organic feed. To stay healthy livestock require nutrients such as amino acids, carbohydrates, fatty acids and vitamins combined with a good balance of proteins, energy and vitamins in their daily diets. While this holds true for most animals, each animal may require different dietary needs. Typically, organic livestock feed can be adjusted to fit the specific nutritional needs of some animals (Cornish *et al.*, 2016).

Comparisons of milk production and feeding ratio from the organic livestock farming to the conventional one, reports show that greater milk production is observed in organic farms with low feed supply to the animals. In addition, organic livestock is observed with a low calf mortality rate, fewer reports of mastitis, and abortions. Moreover, pigs raised under the organic system respond with low respiratory issues and tail wounds. Organic livestock systems with pasture-based and low-input are more socially and environmentally reliable. In contrast, non-organic farms invest in doubling the feed supply, which results in greater extravagance on feed purchase. According to Keifer *et al.,* 2014, organic dairy cow farms have performed better than those conventional ones in terms of economic reliability, long-term sustainability, etc.

Beyond money and ethics, organic livestock farming practices result in numerous environmental benefits. The Organic Trade Association projected that if every livestock farmer in the U.S. converted to organic production, we could eliminate 500 million pounds of harmful pesticides in the environment annually. This pesticide and chemical use in live stocks results in many negative environmental issues. They allow disease resistance to build up in livestock and cause other side effects such as allergies to the animals. The pesticides and chemicals once blown by air during livestock application contaminate the air, water supply and soils. Sometimes these pesticides' harmful effects are encountered for longer periods from the time of application. They may end up finding their way to human food from the contaminated soils and water. This thereby makes natural organic animal rearing practices far better than chemical methods.

Humans have benefited from consuming healthy meat from organic livestock. These animals are believed that they grow up to be healthy and strong as compared to the conventionally reared animals. They possess traits of well-balanced and chemical-free diets with better quality lean meat that lacks unhealthy fats that industrial animals carry. Their products are good for human consumption, as they contain no additives including no preservatives, flavourings or colourings that the current population is avoiding due to their contribution to low health status in their bodies. Though these chemicals keep food fresher for a longer time, the long-term effects of consuming them can cause damage to humankind's health. Consumption of organic livestock products would further ensure that one does not consume genetically modified organisms (GMOs) that are often used to fight issues such as disease, pests or poor weather conditions(Clonan *et al.*, 2015). They have globally raised concerns because of their unknown health effects and possibilities for contamination. These are not found in organically farmed livestock leaving consumers of their products healthy upon consumption. The usage of natural methods in pest and disease control in animals ensures that chemical pesticides that are harmful to humankind are not used.

Organic livestock keeping is beneficial to the environment in that the manure produced and placed in the soils helps in soil maintenance and restoration practices as organic fertilizers. These encourage soil fauna and flora, improving soil formation and structure and creating more stable systems. In turn, nutrient and energy cycling is increased and the retentive abilities of the soil for nutrients and water are enhanced, compensating for the non-use of mineral fertilizers. Such management techniques also play an important role in soil erosion control. The length of time that the soil is exposed to erosive forces is decreased, soil biodiversity is increased, and nutrient losses are reduced, helping to maintain and enhance soil productivity. Farm-derived renewable resources usually compensate crop export of nutrients but it is sometimes necessary to supplement organic soils with potassium, phosphate, calcium, magnesium and trace elements from external sources. The manures from organic livestock are environmentally friendly with negligible side effects to the water sources, surrounding habitats and soil. The machinery used in nonorganic farming releases large amounts of carbon dioxide into the environment .organic farms use natural and chemical-free ways to harvest their livestock, not only in creating a healthier product for human consumption but also for the betterment of the environment.

The potential to preserve biodiversity is promising in organic livestock farming systems than conventional ones. Reduced use of external inputs, enhanced nutrient cycling, precise and active dependence on non-renewable resources have built in a stronger backbone to this system. This would also enable the system to renew all the existing natural resources around, such as water and soil. Organic livestock systems have reportedly contributed low GWP (global warming potential) compared to the conventional model Organic livestock production is often regarded as an important pillar of sustainable rural development. This is because the organic production model generates more positive and sustainable externalities than the conventional model generates. The Organic production system in livestock keeping not only creates an income, but also promotes the conservation of agro-ecosystems from a long-term perspective. Another fascinating thing about organic livestock farming is its active contribution towards building a stronger economy focused on reflecting the livelihood of rural communities.

Lastly, organic farming models have proven more water-efficient in terms of usage and retention on the field, which leads to developing resistance against drought. Moreover, the low input farms experience minimum issues related to soil-land degradation. Many authors and research experts have claimed that organic agro-ecosystems possess greater agro-biodiversity enhancing abilities. As a result, they exert more resilience to various ravaging pests, diseases and climate change in the ecosystem (Clark *et al.*, 2016). Organic livestock production is a means of food production with a large number of rules directed towards a high status of animal welfare, care for the environment, restricted use of medical drugs and the production of a healthy product without residues of pesticides or medical drugs. The International Federation of Organic Agriculture Movements (IFOAM) has formulated the intentions of organic livestock production. The consequences of these rules for the health of the animals were not yet fully anticipated at the time these regulations were made and it has become clear that in some cases that the productivity of such enterprises is beneficial thereby hampering the need to increase the adoption of this production system.

1.8 ANIMAL WELFARE AND ORGANIC ANIMAL HUSBANDRY

Just like human beings, livestock has to be treated specially. The success of any organic livestock farm or enterprise lies in the way animals are treated, the connection they have with the farm owners and the delicacy they obtain from their masters. The world has evolved to a generation where animals ought to obtain protection, insurance and day-to-day review of their conditions and their healthcare. Since animals make part of a larger community of sentient creatures, even believes have their reasons for forgiving them a special place, a place where care is not a factor but their genuine right.

In a society of living beings, animals and more specifically livestock have a special place as compared to plants. While crops survive all by themselves, animals have laid upon humans a set of moral obligations for their survival. Before an animal dies, it is the responsibility of the caregiver to do all he/she can to protect the animal's life. The death of animals in some communities is non-acceptable and can even be termed taboo from the master's angle. Some societies go to the extent of believing in members of the livestock community as their gods and extremely special creatures sent to them by God.

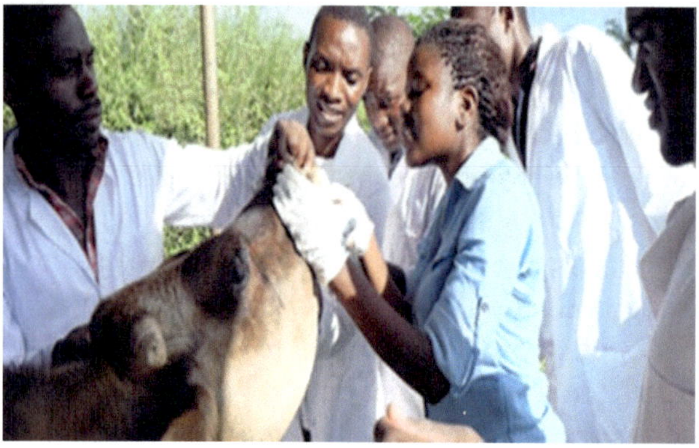

Figure 1.9: Animal treatment is a common part of animal welfare. Organic livestock rearing allows for the use of synthetic medicine.

Source: welttierschutzz

Animal welfare on organic farming goes beyond the art of maintaining animal health. It traverses the aspects of animal freedom and their ability to control what they want to do at what particular time. This implies that animals obtain unlimited access to natural behaviour. Animal welfare is a game of "you do me I do you". It correspondingly leads to huge production improvements awarding the farmer of his kindness to the animal. The improvements in animal welfare over the past years will see organic livestock production boost the world organization for animal health (IOE) has placed a raft of measures and policies to ensure that animal welfare does not march backward but surges forward(Chander *et al.*, 2011). Countries, which rely on organic animal farming, should come up with measures that not only protect animals but also boost their ability in the market. Take the example of Namibia, a country vested in the African continent. Organizations in Namibia came up with animal welfare standards that farmers had to meet for them to supply their products to the international market. The Namibian quality assurance scheme has enacted measures of animal well keeping which have gained the country a lot. Internationally, Namibia is known as one of the countries producing safe meat for consumption from how they emphasize animal welfare. This is complete proof that welfare-oriented products can be more marketable and at the same time build a strong name for the industry.

It is unimportant to note that animal welfare also allows for the use of synthetic medicine for the treatment of animals. Sometimes, we have to accept that diseases keep on evolving and may demand that we react accordingly to save organic farming. New diseases can never be handled by ancient and organic ways of treating animals. It is hence the only exception that synthetic medicine is used in the treatment of livestock to avoid any eventualities. Sometimes, this may result in the livestock losing their organic nature but may as well be the only way possible. As a matter of fact, this step is recommended and may at times be considered a priority.

However, diseases can well be managed if animal husbandry is prioritized based on the following important factors;

Allowing animals with unlimited access to the natural environment rather than in-house control of animals. This allows animals to get and acquire natural nutrients and natural immunity from the environment. It also reduces the prevalence of the spread of new disease infections. The unlimited access should also be controlled having in mind that feeding pastures ought to be maintained.

Improved breeding. When diseases evolve, organic livestock breeds can also evolve. Farmers can rear disease-resistant breeds to reduce the happenings in terms of diseases.

Introducing a well-balanced diet: Many farmers believe that livestock should only feed on grass. That is a misconception, as the grass will provide only a certain variety of nutrients to the animal body. Farmers should instead widen the animal food scope and include items such as green fodder and potato vines just to mention but a few (Celi *et al.*, 2017).

Species appropriate housing. In a bid to maintain animal welfare, housing plays a very important role. The kind of structure organic livestock live in ought to be well hydrated, with the correct kind of temperature. Furthermore, the houses ought to be maintained from the prevalence of pests such as ticks.

1.9 ORGANIC CROPS VERSUS LIVESTOCK PRODUCTS

About 1.8 million producers in at least 160 countries practice organic agriculture globally; production of organically grown food continues to increase steadily every year. Most tropical countries are now exporting organic agricultural products of both crops and livestock. Organic beef from Brazil and Argentina from organic livestock products have particularly taken off exports at the global level while some organic livestock products are yet to take off. Most trade in organic livestock products is however restricted to the European Union and other developed nations in the World. Nevertheless, tropical countries cannot afford to neglect this emerging system of animal production. The demand for organic products globally is very attractive to many farming investors even in developing countries with promising better returns than in conventional production. Organic production is knowledge and management intensive. Producers must be well versed in organic production standards, principles and practices, which require a high degree of knowledge and skill. In organic production of both crops and livestock products it is not simply the final product but the whole production process that must be inspected and approved by the accredited certification bodies in charge of organic products. Organic livestock farming is still evolving, and further research is needed on its products to make it sustainable for the near future (Von *et al.*, 2004).

Efforts are being made to harmonize organic standards for organic crops and livestock products globally. These have coupled with considerable success since most of the organic standards of the EU, the USA, the International Federation of Organic Agriculture Movements (IFOAM) and other regional and national standards show parity, especially those concerned with organic crop products and production processes. However, when it comes to livestock products and production systems standards, there are significant variations in interpretations and requirements in place. For instance, the EU requirements on organic feed and fodder, antibiotics, etc are different by some extents from those of the USA, put across in the National Organic Program (NOP), although the different bodies involved have tasked themselves with the objectives of achieving harmonization in these standards especially in efforts of enhancing trade of these organic products between countries. Their prospects look promising. Moreover, organic livestock production is more demanding than crop production. Organic livestock production relies on organic crop production since, in rearing livestock organically; their feed from fodder crops must also comply with strict organic standards laid down by organic crops authorities. Although organic livestock is, still an emerging promising venture it has yet to overcome several barriers before it becomes popular like the organic crops production enterprises particularly in developing countries. In contrast to organic crops, organic livestock products require good animal health and welfare organic husbandry management practices. Animal rearing is not just a part of the farming system in the field but they are also sentient creatures with fresh and blood and as such deserve special care and moral consideration, unlike crops. Animal management is therefore different from crop management as much as both provide valuable products to humans. Human beings have a moral obligation to treat animals well and to intervene when they get ill for medication before they suffer or die, as this is uncouth. Organic farming principles go further in promoting animal welfare to a large extent than simply avoiding suffering in the hands of humankind. They recommend unlimited access to natural behaviour, which substantially broadens the concept of welfare to livestock. Organic livestock production ensures animal welfare. The rising importance of animal welfare in organic animal production unlike in organic crop production may also get a boost for good products in coming years (Vanhonacker *et al*., 2010). There will thus be recognized benefits for developing countries that can evolve niche markets in higher welfare products globally, and the World Organization for Animal Health (OIE) process for developing baseline standards for animal welfare will assist in the harmonization and equivalency of standards in this area to ensure trade progress among nations.

Cultivation of organic livestock products is fundamental to agricultural crop production as far as it provides the necessary organic matter and nutrients for cultivated land and accordingly contributes towards soil improvements and the development of sustainable agriculture. The wastes from organic livestock as manure aid in enhancing soil fertility for crop production on the farm. Livestock organic products and crop products are highly recommended to lactating women, the old and children, as they are nutritious to their bodies and helps to strengthen their immune systems at large. Organic livestock products are rare to find as compares to organic crop products because of the high inputs requirements in the organic livestock product creation as compared to organic crop product generation.

Organic livestock products and crops have both lower initial investment costs with extra expenses over conventional products farming being organic certification costs and the potential necessity to invest in a better land and special equipment. Despite this, organic products cumulatively do not require the same high capital investments than conventional farming does. Organic farmers avoid the expense that comes with acquiring chemical fertilizers, pesticides and genetically modified seeds and feeds. The investors of organic products from the beginning may not need to borrow as much money from lenders to get started (D'Silva *et al*., 2017). The products are also known to be drought resistant than the conventional products ensuring a continuous supply of these products to humans. The increasing demand for these products and the willingness to pay for these products for health purposes to humankind and concern over the environment suggest that the market for these products continues to make a niche for better returns.

1.10 THE FUTURE OF ORGANIC LIVESTOCK REARING

Organic livestock rearing has today evolved from being outdated to a perspective of the future. Recently, most human diseases are caused as a result of the food we eat or drink whose largest contributors are plants and animal products. Diseases such as cancer are caused by the kind of food we eat. With this being a major concern, organic products are slowly becoming a hot cake in our modern world.

Today, organic milk is more expensive than soda while meat products grow at prices similar to a crypto-currency. The future is not only bright for organic farmers but also promising. However, the environmental dangers facing the earth today may wipe out organic livestock rearing in just but the course of one human lifetime. The high encroachment of desert-like conditions in almost every continent puts livestock at risk of starvation. High temperatures all over the globe inhibit animals from moving from one place to another in search of pastures.

Governmental organizations should take up the role of protecting organic livestock rearing. It is good music for farmers to hear of world superpowers meeting in Paris for climatic policies that reflect largely on organic livestock rearing.

1.11 THREATS FACING THE DEVELOPMENT OF ORGANIC LIVESTOCK FARMING.

Farming as a whole is done based on two factors, food and money. While many people believe in agriculture as the source of food for the entire globe, many others believe in it as a fair source of income. Indeed, most farmers who practice organic livestock farming conduct the same on the basis that they provide clean and healthy products which would not only be relied upon as good food but would also enrich their pockets. Many countries, especially developed ones rely on organic livestock products such as milk and meat from dairies and ranches for food. In the United States of America, organic livestock products are regularly imported and hence earn income to farmers from the origin country (Dippel *et al.*, 2014).

However, the stability of modern market systems can no longer be relied upon. Over the past years, business trends have taught us that one can never be assured of stability in business. Today, market trends can be changed and determined in a matter of seconds depending on the word in the market streets. Factors such as inter-country relations have so many grown roots inside the business and can easily affect farmers dealing with organic livestock. If tariffs and heavy taxes are imposed on farmers and products from specific countries, organic livestock farming may be put in danger. It is clear that over the past years, more so before the dawn of sophisticated technology, agricultural products have been in high demand. However, there is currently no guarantee that this may remain to be so.

Most developing and third-world nations depend on exports to nourish their economy and uplift organic agriculture. Such countries are often ranked as the best on matters concerning agriculture. Although this may act in their favour, farmers from such countries should never be comfortable with international trade. It is common for international markets to demand a completely organic and certified form of agriculture. However, third-world countries are often enriched with a big number of small-scale farmers whose contributions when put together make them strong. Developed countries engaged in organic farming on the other hand are often combined into one sole large-scale production entity. It is easier for a large-scale corporation to prove the organic level of their agricultural products than many organic farmers. The small-scale organic farmers are often hurdled by this and crumpled into begging for market structures.

Market bullying is also a major constraint for organic farmers from developing nations. They are always required to possess a number of requirements before joining the international markets. For instance, the European Union is the largest producer of organic milk, chicken products, organic meat and other products. They are always entitled to supply organic products to their colleague-developed nations. The developing team is always left out because of being bullied out of the market. The developed nations put up many restrictions on the Imports from developing nations. Taking into mind that developed nations have surplus production of organic products, developing nations stand no chance against them (de Jonge *et al.*, 2013). Their only shot is to take extreme steps to make their presence felt in the market. It is absurd that in the same countries, their governments, rather than helping the poor farmers also end up taxing them in extreme ways to gather funds for corrupt practices.

In some cases, there is an upsurge in trading and it occurs that supply comes to exceed the demand for organic agricultural goods. With this, farmers end up getting into a huge loss and more badly, the same emotionally torture them. Organic farmers would end up leaving behind organic agriculture and get involved in other moneymaking platforms. This condition can also lead to a drastic change and huge difference in the pricing of both organic and conventional agricultural products. In fact, this goes to an extent whereby organic agricultural products are used as conventional products. In the past century, organic farming has ended up being unprofitable.

Technology on the other hand has also brought good and bad to the organic agriculture industry. On matters of marketing the use of technology has edged out a huge number of illiterate farmers who rely on and uphold organic farming. In third world countries, there is a notion that agriculture may be put aside for individuals who cannot get access to education and are hence not well educated. For decades ago, people from underdeveloped nations have depended widely on agriculture for food and barter trade on the vast African continent. With there being limited access to education and slow technological advancement, they cannot access their colleagues from developed nations. As trends and agricultural evolution goes on, they are always left behind to cater for their selves. This is a huge threat to organic farming as education is not passed down to others.

In some situations, farmers from third-world countries often face the threat of inadequate supporting infrastructure. State departments and governments are reluctant to enact policies that would protect and provide necessary equipment for the development of organic agriculture. In agricultural India for example, the NPOP adopted in the year 2000 is yet to spring into action 21 years on. In a continent of 1.4 billion people, 17 million of them being farmers, only four agencies are entitled to accreditation. Like as if to add an insult to an injury, the few agencies available only deal with fruits and vegetables, coffee, tea and a variety of spices (Donnik *et al.*, 2016). These agencies are not only inadequate but also prove the inefficiency of some governments in supporting organic agriculture. It is alarming that neither of the agencies deals in organic livestock, a major humiliation.

Farmers may also encounter low produce from organic livestock. In a planet that is currently dominated by climatic glitches, livestock has a hard time adapting to the same. While cows reduce the amount of milk produced, the meat from bulls also depreciates. Not only is this a hazard but also a menace to organic livestock. This can only be turned around through positive climatic views since organic is simply natural.

1.12 PROBLEMS ENCOUNTERED IN DEVELOPING ORGANIC ANIMAL HUSBANDRY

Problems are commonly confused with threats. While threats generally look at the future, problems base their foundations on the present and try to shape up the feature. Indeed, organic animal husbandry is a subject of major interest in our present and current society. The following are some of the problems being encountered in the development of organic animal husbandry.

Diseases

Diseases are one of the major hindrances to all life on earth. It is enlisted as a problem to organic animal husbandry will come as no surprise. The danger is not only habituated in diseases but also how diseases can spread and affect the livestock. Many livestock dies annually as a result of diseases some of which had never occurred in the past. When an animal is ill, several effects occur:

There is a possibility of a spread to other livestock in the herd. In this case, the animal or animals in question are required to be isolated and treated with a specialized veterinary officer.

Danger to humans and other animals from the consumption of sick-livestock products

Jeopardizing trade because of the de-nourishment of livestock. Products may also reduce in quantity and quality reducing the market price of the product at large.

Some of the commonly known diseases in livestock include

- Foot and mouth disease (FMD)
- The swine fever
- The rift valley fever

If a disease is noted in livestock rearing regions, there is a likelihood of a ban in commercial selling of livestock products and more so in the international market. Such adverse effects make disease prevention a major priority for farmers and governments at large. Most governments have turned into creating disease-free zones to curb the spread of diseases. However, livestock disease in most African countries has been poorly managed with many criticisms targeted at government agencies and institutions (Durmic & Blache, 2012).

There is a need to concentrate on animal welfare to promote organic livestock rearing. In our society, much needs to be done to meet the basic standards of animal rearing. Two things need to be prioritized, the animal's health and better quality products. We ought to construct a wall that will distinguish conventional farming and organic farming. Currently, it all seems like organic farming sleeps in the same bed as conventional farming.

Livestock Feeding

The pillar to animal and livestock rearing is the availability of food. As much as it determines the quantity and quality of animal products, it also relates to the health of an animal. Livestock feeding has sparked debate at the international level with modern developed countries putting their attention into this. In the United States, for example, the USDA has come up with a several policies that control the use of pastures for livestock feeding. This move strengthens the use of pastures and controlled its growth.

In the policies, animals are entitled to 30% dry matter on grazing pasture, which should be done under strict management. Compare this to African countries where pastoralism is still a major subject. In this scenario, an individual can graze his/her livestock in whatever amount of time and in whatever way as long as the ground is not bare. As much as it is understandable that animals obtain food for them to eat and survive on they also constantly need them. These policies concerning grazing make control a subject and at the same time make pastures available all year through.

Insufficient Knowledge

Organic production and livestock rearing are more of an educative subject than a regular practice. A proper running of an entity, business or organization requires vivid learning of the subject and so is organic livestock rearing. It is absurd that a vast number of farmers have limited knowledge of organic farming. First and foremost, most livestock farmers in their childhood ages are made to believe that one can practice organic livestock rearing rather well without education. In other regions, there is no platform for educating individuals interested in organic livestock rearing.

There are key sections where people need to have an in-depth understanding of organic livestock rearing (Vaarst *et al.*, 2012). This includes the principles of organic livestock rearing, its standards, the production practices in livestock farming and the various requirements for certification with farming agencies. With the growth of the internet, it is far more important that all farmers be educated on the importance of using technology in embracing organic livestock rearing.

It is much harder for education to be embraced in extremely poor regions where people depend on organic livestock rearing for a living. There is a higher number of illiterate individuals in such regions, which makes it hard for people to access education in the available languages.

Small Farms

With the majority of organic livestock farmers coming from third-world countries, most of them rely on extremely small portions of land in rearing their animals. It is important to note that landless animal farming is not titled under organic livestock farming. This is because small farms are not favourable for organic livestock rearing. Small-scale rearing of livestock also comes with many challenges. For instance, in milk production regions where small-scale farming is prevalent, the farmers are known to indulge in milk contamination, dilution and traceability.

Such requires policies and technical measures to come up with solutions to small-scale farming.

Sanitary

They say cleanliness is next to godliness. For maximum production and quality livestock products, cleanliness is key. Most importing countries gave a raft of regulations regarding the sanitary levels in the phase of organic livestock rearing. It is only a few developed countries, which can meet sanitary requirements for organic livestock products. Sanitization is possible through personal will from farmers. Farmers from tropical countries are emphasized to enhance hygiene to ensure maximum production of animal produce (Thornton, 2010).

Chapter 2

Factors Influencing Organic Livestock

CONTENTS

2.1 INTRODUCTION

Over the years, there has been remarkable improvement and increase in the numbers of farms practising organic farming. This is due to the ever-increasing demand-supply for food more so for animal products. Farmers have been able to venture more into livestock farming and incorporating various technologies to ensure its success. Improved systems have been incorporated so that livestock farming can facilitate increased profits for the farmers and increase food safety. This farming is also done to ensure that all ethical concerns are taken care of. Farming is also done in a manner that facilitates environmental protection. For these reasons, organic livestock has shifted its level from being despicable to a dependable source of food. However, there is a challenge in defining the impacts of systems used in organic livestock farming in ensuring that the general food system is sustainable. The lack of established impacts is due to the lack of definitive characteristics of organic livestock farms and the poor definition of the contexts of such farms. This is coupled up with the fact that several limitations are hindering the development of this sector. This has influenced the future perspectives of the sector (Escribano, 2016).

In the organic livestock sector, little research has been done to analyze and address some of the challenges it faces in helping achieve food sustainability. This will cause a limitation in integrating new strategies that will help improve the sector. Research is underway to improve knowledge of organic livestock farming. By doing this, farmers and researchers can get more insights and different points of view they can use to help relevant stakeholders tackle some of the challenges in organic farming. They will be able to focus on ensuring that the farming practices will ensure the sustainability of the farms. It will positively impact the food system by ensuring increased food production, distribution and security. The new knowledge will help farmers and relevant stakeholders in making informed decisions at different levels. This information is also useful for policymakers as they will be able to develop strategies that will ensure that organic livestock farms will be successful.

Several factors are influencing the success of organic livestock farms. They are as follows:

2.2 ESTABLISHING OF ORGANIC FARMS

With the implementation of organic farms into the food system, there has been a challenge in successful implementation. More research has been done to ensure a smooth implementation. Studies have been able to show the impacts of transitioning from livestock farming to organic system farming. Research indicates that for there to be a successful implementation, there is a need for a good livestock farming structure. The feasibility of the farm is also dependent on the context of the farm. For example, in the Mediterranean basin and Europe, there are ruminant-based pasture farms that carry out livestock farming. With these farms, it is quite easy to convert them into organic livestock farms. This is because the farming systems used to bear some resemblance to those used in organic livestock farming.

However, there is will be a limitation in the conversion of the breeds of animals reared. The challenge is made more difficult in farms that carry use intensive agricultural production systems. Some of the animals in this category include dairy cattle breeds, poultry and pigs. Some of the farmers will face challenges such as the availability of breeds of such animals to facilitate organic farming (Tallentire *et al.*, 2019). For farmers who opt to convert to organic livestock farming, there are several factors to consider. They include the motives of the farmer, the perspectives of the farmer and the status of the farm. For farmers who keep ruminants, they need to have access to land so the farming systems to be well implemented.

For farmers who keep monogastric animals, they are more flexible as they do not require much land to carry out organic farming. However, for these farmers, it is difficult to convert farms to use organic farming systems. This is because farmers will not be able to afford the organic feeds as they are expensive.

Concerned boards need to take necessary measures so that the organic feeds are sold to farmers at a cheaper price. Farmers rearing pigs using the free-range system of production have been known to successfully convert into organic farming. This is because the models utilized in the rearing of swine have been modified to suit it. The success stories have been recorded in countries such as Spain. The ecosystem in the southern parts of Spain is known to positively influence the successful implementation of organic farming. Farmers will need to be very keen at the weaning stage of rearing swine. This is because several veterinary services will be required to ensure the survival of the animals.

Apart from the species of the animal, the farm structure will also play a role in ensuring the successful implementation of organic farming systems. For farmers who carry out mixed farming, they use systems that are of high resilience. The mixed farming production system has been designed in a manner that allows an easy transition into organic farming. A good example is when mixed farmers graze both heifers and pigs on the same piece of land. It is beneficial as it is a way of controlling pests known to harm heifers. How pigs are inclined in a manner that facilitates rooting. This has been handled in a manner that makes farming activities such as ploughing much easier.

Research shows that keeping poultry in significantly large flocks is more beneficial in that the poultry will play the role of weeders. They are also good at controlling certain crop pests. This system of farming is applicable because different species of livestock have different feeding habits. The farmers can make use of that information to their advantage. They can also use feeding habits in the management of crop pests and diseases. This method is mostly used in rural areas in ensuring livestock farming sustainability.

In as much as the conversion of livestock farming to organic livestock farming is beneficial, there are some impacts incurred when using this system of farming. The impacts vary due to the variety of factors in different ecosystems. Some of these factors include the topography of the land, climate experienced in the region, the environmental ecosystem, government policies, standards of certification, farming systems used, ease of access to organic farming products, prices of organic inputs, demand for agricultural products and marketing strategies used(Smit *et al*., 2017). Researchers have tried to establish how these factors impact the transition of livestock farms into organic livestock farming.

There has been limited knowledge on how different farms that make use of conventional farming systems can transition to organic farming. More research has been done to establish the ease of transition of a variety of conventional farming systems. The study was conducted for dairy cattle and goats. Researchers need to establish the challenges facing the implementation of organic farming, the motivation for implementation and the farmer's objectives. When conducting a study, a multidisciplinary evaluation should be done to gauge the sustainability of the farm and the likelihood of successful organic farming. After the evaluation is done, a SWOT analysis is done. The competitiveness of the farm's understudy needs to be evaluated for researchers to get accurate results.

With regards to the rearing of beef cattle, there are many uncertainties in the conversion of livestock farming to organic farming. There are also some arguments against the use of organic livestock farming. This has resulted in many studies being done to put to rest some of the arguments. The studies involved relative comparisons being done between organic farming systems and conventional farming systems. The studies were done in a fixed duration of time; a year's study to study the sustainability of the farms over long production cycles. Most of the farms under the study were under the Common Agricultural Policy. Results of the studies show that farmers who practised organic beef farming get relatively poor economic results compared to those who use conventional farming systems. It also showed that organic farms tend to rely on price premiums and subsidies. When compared to conventional livestock farmers, organic livestock farmers tend to incur higher costs of production. This is because most of the organic feeds are expensive and a significant amount of money is paid when the farm is transitioning to organic farming.

A study was conducted to evaluate the success of adopting organic farming systems by conventional suckler farms. The study involved conducting an analysis being done to evaluate the effects of conversion on-farm production, its economy and the environment. The adoption of the organic farming system has contributed to the banning of chemical inputs such as fertilizers. This has resulted in a decline in meat production which has caused reduced income for farmers. Despite the decline in production, an improvement has been noted. There is a significant increase in the selling price of meat. This will cushion farmers allowing them to get some profit despite the decline in agricultural production. As a means of preventing farm losses, most farmers opted to minimize the use of organic inputs. This helps them ensure the sustainability of the farm.

Studies were also done to evaluate the impact of the adoption of an organic system of farming in milk production. It was noted that organic farming systems have positively impacted milk production. When milk production is evaluated per animal then a noted decline will be in milk production. This may be viewed as a disadvantage. However, it is not. In the initial stages, the milk production per animal is lower. However, the use of organic inputs has been known to increase the longevity of the animal allowing it to produce more milk (Escribano *et al.*, 2019). Organic farming has reduced the need for farmers to heavily invest in the rearing of heifers. Therefore, organic farming has made farming to be more economical for farmers.

In conventional farms, it was noted that in as much as the animal production level is high, it has resulted in an increased incidence of the animal suffering from certain health complications. It was noted that most of these animals tend to have a somatic cell count higher than the normal somatic cell count. They also tend to suffer from diseases such as mastitis. Cattle under this system tend to be less tolerant to heat and have a tendency of having reduced fertility. For fertility to be maintained, milk production of the animal will have to be minimized. When the production is reduced, the farmer can reduce the number of monosaccharides and glucose given to the animal. This is beneficial as it reduces the likelihood of animals suffering from certain disorders. However, it means the animals will need more feeds to supplement the lack of non-structural carbohydrates. This is not the case for animals under organic farming systems.

Therefore, organic farming structures have resulted in increased milk production while a reasonable amount of feeds is given to the animal. It also improves the health of the animal while ensuring that it has a long production life. This meets the objectives set by TP Organics. The organization aims at conducting research and developing innovations that will facilitate the organic farming and production of organic products.

Over the years, the price of organic milk attained stability. However, when the demand for milk is low, the best approach for farmers will be to allow calves to suckle their mothers. This prevents losses due to milk spoilage. It is also beneficial as it will improve the health of the calves and farmers to not have to purchase many feeds for the calves. For farmers who have adopted this strategy, they have discovered that calves that suckle their mothers tend to have a higher weaning weight compared to those that do not suckle. When looking at the economic benefit of the farming systems, it was discovered that cattle under organic farming systems yield more economic advantages compared to those under the conventional farming system.

Not only is organic farming applicable for ruminants but also other animals. Research has been done to evaluate the successful implementation of organic farming in the rearing of animals such as guinea pigs and rabbits. In the case of rabbits, research indicates that the success of organic farming is dependent on the genes of the animal and the system of production (Sundrum, 2012). When the effects of the conventional and organic system of farming are compared it was discovered that the daily gain for rabbits under the conventional farming system is higher than those under the organic farming system.

2.3 REGULATION OF CERTIFICATION OF BODIES

Like other agricultural sectors, bodies are controlling the production and sale of livestock products. The government has also put in place certain legislations that will control the production of livestock products. The regulations outline the type of animal breeds to be reared, the type of agro-ecosystem management, the agricultural systems to be used and the marketing and feeding structures to be used. Most of these regulations are used in organic farms. By so doing, they can increase the likelihood of the farms succeeding. It will also help farmers get a new perspective on how to handle challenges in organic farming. A good example is the case of the variation in the amount and quantity of different types of feeds that are independent of the amount of milk produced in the north and south of Germany. The two variables were different in the two regions. This was attributed to the variation in several factors. Some of these factors include the ecosystems in which these farms are located and the demand for these agricultural products.

In the establishment of the bodies in this sector, many requirements have to be met. This applies to both public and private bodies. They are the ones who decide whether there are some exceptions to be made for some farms in following the regulations depending on the farm level. This is why all the boards meet and make a consensus on the regulations and policies. Among the requirement is that the farmers need to ensure that their farms have been certified. A certification is a form of a permit for them to practice organic livestock farming. The challenge that most farmers face is that in most cases the cost of certification is high. This makes most farmers (more so small-scale farmers) unable to afford certification. This renders them unable to practice organic livestock farming significantly impacting food sustainability and security (Entz *et al.*, 2009).

This has created the need for needed efforts to be put in place to enable farmers to get the needed certification. This includes boards being put in place that will help in the marketing of organic products globally. There are also some agreements that have been made between the American and European standards through the Participatory Guarantee Systems that will make certifications of farms more affordable. The other challenge faced by farmers is the lack of regulations that control the marketing of organic products. Most of the regulations and standards in the public sector do not cover the marketing of such products. This makes it hard for the relevant stakeholders to evaluate the contribution of organic livestock farming in ensuring food sustainability.

2.4 AVAILABILITY OF PUBLIC SUBSIDIES

In as much as organic farms record low production levels, the positive impacts it has on the environment should encourage relevant stakeholders to champion the implementation of policies put in place by the EU's CAP. The implementation of these policies more so in the rural areas will play a significant role in ensuring the conversation of natural ecosystems. Most of the policies put in place embrace the use of a management structure that is more natural and eco-friendly. This is crucial as it will ensure the development of rural areas. It will also encourage farmers in these areas to invest heavily in agriculture. By so doing they are acknowledging the role played by advanced livestock farming systems in preserving cultural and environmental heritage.

2.5 ANIMAL DIETETICS

Good nutrition is important for both human beings and animals. This is why it plays a significant role in livestock production more so when organic farming systems are utilized. Different feeding strategies are used by farmers to ensure the good health of the animals. Research shows that in organic farming the feeding strategy used will determine the level of animal production and the profit incurred by farmers. Plenty of research has been done to obtain relevant information that will make farmers make more informed decisions. Available information on organic farming and the process of conversion to organic farming has served farmers.

For farmers to ensure that their animals are healthy, they need to adopt farming strategies compatible with those for organic farming. For instance, animals need to be given an organic pasture. More research is done to evaluate the need for nutritional supplements. They also help farmers handle some of the feeding challenges. Most farmers practising organic farming face two major challenges; the high price of organic animal feeds and their unavailability. This is worsened by the fact that most farms are highly dependent on organic animal feeds. This is a result of the dissociation present between livestock and plants. These challenges make it hard for most farms to completely implement organic farming systems (Sutherland *et al.*, 2013). The high prices of these feed also contribute to the poor adoption of organic farming. It makes them unable to access the feeds. The challenges present a major setback for organic livestock farming.

The effects of these challenges have been felt by organic farmers. They have to look for alternatives to supplement the animal's diet; some of them incur losses and are unable to successfully incorporate organic farming. These effects are felt in both dairy and beef farms. With the identification of the challenges, it is easier for researchers and the relevant stakeholders to come up with effective solutions to the problems.

Among the challenges is the unavailability of organic feedstuffs. One of the strategies adopted to deal with this challenge is the use of locally available agricultural by-products. This is beneficial and reliable for farmers as they can easily access these products. They are also economical as they are sold at an affordable price. The use of agricultural by-products is also environment friendly as these products are in a biodegradable format conserving the environment. This is also advantageous for individuals producing and selling organic by-products as the use of these products by organic livestock farmers creates a market for them.

The EU's CAP has stipulated some laws that require farmers to make use of more natural remedies. They have laws that limit the use of feed additives. Farmers are forces to use other sources of minerals such as the soil. The problem with using soil as a source of minerals for animal nutrition is that there are regions where the soil mineral content is very low. This results in animals suffering from different kinds of deficiency diseases. Dairy farmers will be severely affected by the inadequacy of these minerals. This is because dairy cattle need a highly nutritious diet to facilitate high milk production. This creates the need for the creation of feeds that will meet the nutritional demands. This includes using crops that are rich in minerals.

Some of the regulations put in place require farmers to rear animals that are fully organic. Before the regulations were fully adopted research was done to establish the impact of organic feeds on the energy stability in dairy cattle. This required blood samples for different cattle from several farms. The results show that there are minimal negative impacts of organic feeds on animal metabolism. There are areas where the feeds are naturally available but the quality is poor. A good example is in the Mediterranean basin where forage is plenty but the quality may not meet the required standards. This may require farmers to supplement the missing nutrients at a certain frequency.

The availability of such feed supplements and additives is also poor. This is because the cost of producing organic products is relatively high. This is because the processes involved in converting feeds to organic feeds are more intense compared to the conventional production process. There are some uncertainties in the profit generated as a result of making use of organic production lines. In addition to that, companies under organic lines of production would need more workers to handle the workload. This will result in decreased company profits. This creates the need for definitive guidelines on the incorporation and utilization of feed additives to be defined by the legislation (Sorge *et al.*, 2019).

Good dietary management is needed to ensure the optimum health of livestock that will result in improved productivity levels. Good animal health is also resourceful in minimizing the chances of transmission of certain livestock diseases. For these reasons, the relevant stakeholders should have talks with conventional companies. These companies are known to produce feed additives that are per the regulations outlined by the EU. The additives are known to assist in rumen perturbation and minimize the utilization of antibiotics. This has resulted in reduced levels of pollution in the environment. The decrease in the use of antibiotics has contributed to the decrease in animal health complications. By working with conventional companies, there would be the production of animal feeds that would improve the sustainability and efficiency of the livestock sector. Policymakers could establish laws that control the use of organic minerals to reduce the chances of environmental exploitation. These additives may have restricted amounts of fat supplements, organic minerals, essential oils and organic acids.

2.6 PRESENCE OF LIVESTOCK DISEASES

The total implementation of organic farming has faced several challenges among them the presence of livestock diseases. In organic livestock farming, good health in livestock is needed to ensure that the animal products will get a good market. The presence of viral, bacterial and zoonotic diseases poses a threat to organic livestock farming and trade. Diseases such as swine fever and foot and mouth disease have causes a restriction in the exportation of organic livestock products to other countries. Relevant stakeholders need to enforce necessary measures that will ensure such diseases are controlled. The actions taken may vary depending on the prevalence of the disease in a certain country.

In India, foot and mouth disease is known to cause severe effects on livestock farming. The government has taken the initiative to control the spread of this disease before it causes severe damages to livestock production. Some of the measures taken include the creation of disease-free zones when organic livestock farming can be carried out. The government has also facilitated free animal health services to help in controlling the spread of the disease. However, this has not been effective as many farmers in rural areas have claimed that they have not accessed these services. The approach used in organic livestock farming is to ensure animals are healthy and the spread of diseases is minimized. This is achieved through good management strategies and practices. This is advantageous as it results in good animal welfare and the production system is efficient.

2.7 ANIMAL WELFARE, HEALTH AND TECHNICAL MANAGEMENT

Over the years there has been a notable increase in the adoption of agricultural methods of farming on a global scale. Veterinary officers need to have more knowledge on organic treatment options to be used on these kinds of livestock. The challenge is there is limited knowledge with regards to the management of animals under this system. Decisions regarding animal health are made to ensure the longevity and sustainability of the animal. The lack of knowledge makes the process much difficult. This also affects other areas such as the environment, human and soil health. Organic farming calls for the utilization of minimal amounts of chemicals as possible. Therefore, the objective of most organic farming projects is to ensure that they uphold good animal health while using minimal amounts of medicines (Souillard *et al.*, 2019).

In accordance with regulations laid out by the EC, farmers practising organic livestock farming are required to use phytotherapeutic products in treating their animals. The challenge is that the availability of phytotherapeutic is limited. Knowledge on the use of phytotherapeutic among veterinary officers is limited. This creates a problem for the farmers and the veterinary officers.

There is a strong relationship between the welfare and health of livestock. Veterinary doctors handling organic livestock are tasked with the role of ensuring the animal's social, mental and physical well-being is maintained. They also strive to ensure that animals are less likely to suffer from livestock illnesses.

To ensure the animal's optimum welfare and good health is a difficult task. There is a need to utilize production systems that will minimize the incidence of animals contracting livestock diseases. The farming system used will determine the level to which the animals get in contact with pests and disease-causing microorganisms. A good example is when a free-range production system is used. In this case, animals are very likely to suffer from diseases. Farmers rearing pigs opt to use the outdoor housing method. The method has proven to yield more positive results. It enables the animals to easily access grass and other feeds. They also get good access to roughage, enhancing digestion. This method of farming also reduces aggression in pigs solving some of the challenges faced in the production of piglets.

Generally, the use of an organic farming system is better when compared to conventional methods as it ensures good animal health and welfare. The challenge is that more commitment is required by farmers to ensure animal welfare is maintained. Researchers are trying to establish the relationship between general organic farming principles and animal welfare. They also look into some of the advancements needed to improve organic livestock farming systems. Good welfare and health of livestock involve good parasite control. Organic livestock farming aims at ensuring good parasitic control and nutritional management. This is because it was realized that when different species of animals were analyzed, the level of nutrition was poor and some had intestinal parasites. These challenges were evident in rearing organic goats.

Researchers came to realize that some of the fatalities in organic livestock farming are experienced at the initial stages. Over time, it was discovered that cows were less likely to suffer from mastitis, they had fewer loads of ectoparasites and reduced cases of calf mortality were recorded. However, when a comparison is made between conventional farms and organic farms it was discovered that livestock in organic farms experiences more prevalence of intestinal parasites (Escribano, 2016). Despite the drawbacks, there is hope for organic farmers as animals have been discovered to have minimal parasitic resistance which allows researchers to develop new strategies to improve animal health.

Though organic livestock has shown to have a high prevalence of intestinal parasites different results were obtained when the animals were tested for bacterial resistance. Organic livestock animals had a higher bacterial resistance when compared to conventional livestock. There is also a variation in the treating of livestock. When conventional livestock are compared to organic livestock it was discovered that conventionally reared livestock are more likely to use veterinary medicines when compared to organic livestock. The same case applies to using animal vaccines. This has been regarded to contribute to the variation in somatic cell count between the two types of animals.

The prevalence of parasites in monogastric animals has also been analyzed. This was handled by the COREPIG project. This project was established by Europe to help minimize the spread of parasites and diseases in organically reared pigs. Therefore, control of pests and diseases is the major objective of the project. The project has funded research seeking to provide more information on organic swine production. An analysis is done of the status of swine production and challenges faced by farmers during production. They are different production systems used in the rearing of pigs; some use outdoor management while others utilize concrete runs. There are other farmers who use a combination of different systems. Different breeds of pigs are reared by farmers. In as much as conversion into organic farming is beneficial, there is still the challenge of the prevalence of pests in these animals.

When analyzed pigs reared under organic farming are said to have behavioural freedom however, they are more likely to be negatively affected by changes in climate and infestation by parasites. Therefore like ruminants, the issue of good animal health and a high level of production is a challenge. In pigs, it is a unique situation in that there is a potential risk in public health depending on the meat supply. A study was conducted by the EU on the health and well-being of piglets as they are weaning or suckling. In this study certain parameters are outlined by the Welfare Quality Protocol. They have been used to evaluate some of the problems encountered in these farms (Sossidou *et al.*, 2015). They also evaluate the challenges farmers face in keeping piglets. From the study, it was established that some of the challenges faced in using organic farming are dependent on the type of organic farming system used.

Research shows that in pig farming, the organic system yields better results compared to conventional farming systems. This is evident in the minimum numbers recorded pigs having disorders. For those under the organic system, they are less likely to suffer from disorders, hernias and abscess. Research also indicates that organic pigs tend to suffer from certain diseases such as white spot livers and lesions on the joints. In as much as this is so, it was noted that organic pigs tend to consume lesser amounts of antibiotics. When both systems are compared, there is no difference between their mortality rates.

Some measures that been to deal with the challenge of disease prevalence in animals in the organic system of farming. Developments are being made to controlling the spread of pests. Some of the strategies farmers are being encouraged to take up include rotational grazing. The specialization of organic farms also helps in the use of effective management practices useful in ensuring proper animal health.

Organic agriculture has also been applied in aquaculture. However, it has been established that there several setbacks in using both conventional and organic systems. This is due to the influence of other factors such as quality of water, control of fish predators and availability of feeds for fish. Researchers have concluded that organic farming principles may not allow the adoption of open net-pen fish farming.

2.8 DEMAND AND MARKETING OF ORGANIC PRODUCTS

There is a growing acceptance of organic products among people. It is said to be more intensive compared to other farming systems. This is because it has positively impacted the environment by contributing to the increase in soil fertility. It has also brought about improved quality of pastures and farming biodiversity. It is also environmentally friendly as it has caused minimum environmental impacts with regards to the utilization of land. This system of farming also requires minimum energy in the creation of agricultural processing. The other advantage of using the organic farming system is that there is no competition for food between human beings and animals. This increases the level of food security. This is why most authorities are encouraging farmers to convert to organic farming. Statistics show that one-third of cereals produced globally is used in feeding livestock. Two-fifths of these grains are used in feeding ruminants such as cattle.

For the successful adoption of organic farming, better marketing strategies need to be put in place. This will ensure that consumers are aware of new organic products. The challenge is little efforts are is being used to ensure proper marketing of organic products. This is said to be the reason why consumers are not aware of organic products. The lack of knowledge of organic products has contributed to the poor demand for such products. Most of the organic products are sold at a higher price compared to other agricultural products. This has also contributed to the low demand for organic products. This poses a challenge to farmers as they will be unable to sell their products and if they get to sell them the price will not be able to cover the cost of production. For organic livestock farmers, they have to deal with the challenge of the unavailability of organic livestock feeds. When the feeds are found, they are sold at extremely high prices. In as much as there are available feed substitutes, it still discourages farmers from practising organic farming. Relevant stakeholders need to develop a measure that will enable farmers to access organic feeds. The low demand for organic products has been attributed to the lack of adequate knowledge on organic farming, lack of awareness and the preference by consumers to purchase products sold at a cheaper price. A good example is a market for organically reared cattle. For some time, there was no market for organically weaned calves because the demand was low (Gunnarsson & Mie, 2018).

Some farmers have been able to sell their organic products. The lack of well-defined marketing channels has resulted in the variation of prices between organically and conventionally produced livestock products. This has resulted in high prices of organic products contributing to the minimum amounts of agricultural on supermarket shelves. In a bid to ensure no fluctuations in livestock products there is a need for proper distribution of products.

In the marketing and selling of beef products, it was discovered that the demand for organic beef products is lower because consumers are unable to differentiate between conventional and organic beef products. One way of ensuring increased demand for organic beef products is to ensure that they are of high quality. The challenge is that it is quite difficult to establish characteristics of quality meat products. This is needed when evaluating the advantages of conventional and organic meat products by looking at the nutritive content and the health status. Other factors that contribute to the low demand for organic meat are the variation in consumer behaviours, inadequate knowledge on organic farms and lack of awareness of organic products by consumers. Many strategies are being developed in solving the poor demand for these products. Some of the strategies include the exportation of these organic products. However, with the exportation of organic meat products, the relevant stakeholders need to be able to cater for transportation costs as well as the storing of products.

The most appropriate strategy would be the creation of markets specifically for organic products. These markets need to study consumer behaviour as well as market trends. They need to know the type of products consumers want. They also evaluate if the population is aware of the existence of agricultural products. Are consumers able to differentiate between conventional and organic meat products? What percentage of the population knows the existence of certain agricultural products? Are organic products sustainable to meet public demand? These are some of the questions market researchers seek to answer. They also need to know whether the principles utilized in organic farming are compatible with consumer values. Market researchers know that most consumers will purchase products known to be of high quality. Therefore, farmers needs to ensure organic products meet consumer quality standards. It is also important to ensure the proper distribution of organic beef products (Grunert *et al*., 2018). The agri-food system is tasked with ensuring the proper distribution of organic products. They also need to ensure that the channels used in marketing these products are relevant and efficient. This system is known to make use of conventional marketing channels to market organic products. It has proven to be effective as more consumers are more aware of the presence of organic products.

Chapter 3

Organic Livestock Feed and Their Benefits

CONTENTS

3.1 INTRODUCTION

During the first revolutionary ages, the human race used to depend on greens as a source of food more than they depended on animal products as a source of food. As time went by, they slowly adapted to the feeding of animals, not their products. With each advancement in the revolutionary ages, the human race evolved, so did their way of life change too. It came to a point where some of them opted for the rearing of livestock within their surroundings, yet before all animals were wild to them.

The rearing of animals became a part of human beings day to day lives and a priority to them as much as crop growing was a priority to them. The humans had to look for feeds for their newly made friends, water among other basic needs for the survival of livestock.

Back then, animal rearing was purely traditional. For instance, if an animal got injured or sick, its wound or illness would be treated using traditional herbal medicine. If animal health was to be improved, the human race would feed them in the best type of animal feeds that guaranteed the rearing of animals in good health. For animals whose product was the core for their rearing, farmers used to feed them on feeds that ensured the animals were produced in high quantities at high rates (Honkanen *et al.*, 2009).

3.2 A RECAP ON ORGANIC LIVESTOCK FARMING

As time went by, the techniques used in the rearing of livestock advanced. Each new revolutionary age had a new way of improving the quality and quantity of animals and their products.

In our current age, the most common way of improving the quality and quantity of animals and their produce is by the use of artificial components. The artificial components include the use of artificially engineered feeds for animals to feed on; artificial medicine to treat animal ailments; and even the production of new animal varieties with the help of artificial inputs.

The main disadvantage with our current ways of improving the quality and quantity of livestock and their produce is that; the artificial ways are causing many negative effects both on the environment and on the health of the consumers of animal products. This is one of the many reasons why individuals are opting to go back to the traditional ways of rearing livestock.

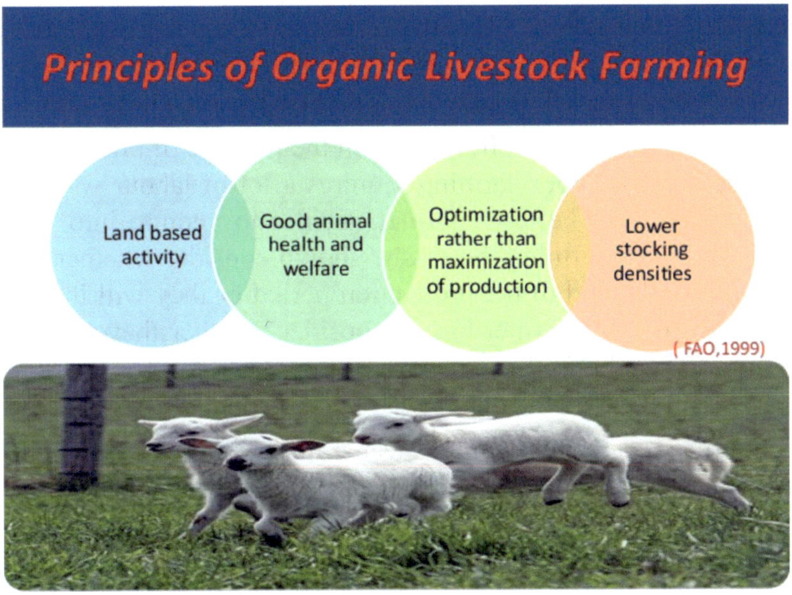

Figure 3.1: An image showing some principles of Organic Livestock Farming.

Source:https://www.slideshare.net/SheikhShubeena1/organic-livestock-farming-95693347

Organic Livestock Farming, as the name suggests, is the use of organic supplements to boost the health of a livestock animal, its production rate and the quantity of its produce.

In most nations, organic livestock farming has become a nationally recognized way of farming. Actually, some nations do offer a certificate for interested parties.

However, organic farming has its challenges. Before on decides to venturing into pure organic livestock farming, they ought to put into consideration a few factors.

One of the factors is the determination of whether there is an easily accessible market that would purchase the products from the farmers. This factor is so critical because, not all markets are ready to pay an organic premium for livestock products since the quantity of organic produce is low compared to that artificiality engineered animals' products; however, the quality of organic produce is of high quality (Ivemeyer *et al.*, 2012).

The second factor to consider is the proximity to a processing unit. Most organic livestock farming is practised in remote areas; hence the availability of a processing unit is low. The proximity is an issue because most of the animal products ought to be processed immediately after they have been produced from the animal since most of them are known to be perishable.

The third factor is the profit rates from the practice of organic livestock farming. Organic livestock farming requires a lot of labour which can be costly to achieve; so, before a farmer decides to venture into the world of organic livestock farming, he or she should consider whether they will make a profit at the end of the production period or they will incur losses. Still, under profits, the farmer has to consider whether they can fully the production of animal's feeds; for example, the consideration of the purchase of a piece of land on which animal pasture is to be grown.

The fourth and last factor is the consideration of the cost to be incurred in the entire farming of livestock. One, the farmer has to consider the availability of the labour force. Should there be enough labour for the entire production process to be a success; are the equipment required for organic livestock farming available; will the farmer produce their own organic livestock feed or will they purchase; among other considerations to be evaluated.

As stated earlier, in most nations, the practice of organic livestock farming does require a certificate to be assigned to the farmer for them to be nationally approved and recognized. The certification helps in situations whereby the country is a need of organic products; all the government has to do is to conduct one of the certified organic livestock farmers and place an order for organic produce.

Before a certificate is assigned, several factors have to be considered. One of the factors is the availability of a big enough piece of land. The land is so important in Organic Livestock Farming for every tiny activity in the farm ought to be purely organic. With adequate land, animals will be able to graze on organic feeds; the spreading of manure for further organic production of organic feeds will be easy; animals will have access to clean and fresh air for they have enough space for movement from one place to another.

The second factor is that a farmer ought to select the perfect breed of livestock. Not all livestock animals survive in just organic farms some thrive well in farms that are managed with the use of artificial components. Livestock breed needs to be determined before the assignment of certificate for it ensures that the animal breed guarantees the survival of the farming practice and that losses will not be incurred in any way. The third factor to be considered before a farmer is assigned a certificate for practising Organic

Livestock Farming is that; the animals have to be treated with humanity just as human beings treat each other. Any discrimination of an animal or making changes to the natural ways of the animal's way of life ought to be eliminated in organic Livestock Farming. Activities such as taming animals into cages; limiting the natural movement of domestic animals within a certain environment should be avoided; mythical activities such as tail docking, beak trimming to aid in the good growth of animals should be discouraged (Jespersen *et al.*, 2017). Basically, a farmer ought to guarantee that the livestock will be reared in the most natural way allowing it to grow in its natural form with no changes being on its structure, and if necessary, only minimal changes. If the environment cannot allow for natural rearing of Livestock, the farmer ought to do everything within their power to ensure that the environment is designed in a way that allows the animals to grow naturally.

As the name suggests, organic livestock farming should be purely organic. The drugs administered to the animals should be organic; meaning the use of herbal drugs should be employed. The feeds fed on livestock should be organic meaning that they ought to have been grown with the incorporation of organic ways of crop production. As stated earlier, the environment should be more natural than artificial setups. Artificial medicines, use of artificial farm inputs in the production of animal feeds, should totally be discouraged.

3.3 FEEDS

Organic livestock feeds, just like any organic food basically means animal feeds that have been produced using organic methods of farming. The production of organic feeds means that; cycling of organic waste materials will be employed; the balance in ecology will be highly considered during the entire production of organic feeds and that biodiversity will be conserved.

With Organic Livestock Feeds, farmers are assured of improved health of their livestock hence a guarantee that their animals will not be prone to common diseases for their immune system has been strengthened and they are resistant to microorganisms introduced to them by pests.

Organic livestock feeds should be free of any artificial food additives, if they are to be processed, they ought to be processed using few artificial methods, and the addition of modified ingredients should be avoided completely. The use of pesticides and herbicides in the production of organic food additives should be discouraged.

Figure 3.2: An image showing cultivated organic Livestock feeds (Organic Fodder).

Source: https://infonet-biovision.org/AnimalHealth/Fodder-production

With organic Livestock feeds, there are beliefs that such feeds are safe, with high nutritional value hence boosting the animal's immunity among other benefits. Organic feeds are said to provide an animal with the necessary nutrients that boost its health and prevent alteration to an animal's hormone system. Organic animal feeds are said to contain minimal to zero chemical contents, giving them a higher advantage over chemically produced organic feeds. Anti-nutrient contents are also believed to be less in organic feeds compared to their percentage in artificially produced animal feeds.

Organic feeds have been discovered to have minimal percentages of pesticide residue. Pesticide residue is referred to as the amount of pesticide that remains in a crop after the pesticide has been applied. Pesticide residue is minimal in organic feeds because the use of pesticides is totally discouraged in their production.

Basically, the use of organic feeds is guided by the principle that healthy animals mean healthy people; meaning that, individuals who feed on products from healthy animals, will end up being healthy; and for the animals to be healthy, they must be fed on healthy feeds. Healthy feeds not only improve the growth and health of livestock but are also important in the breeding process (Kleter *et al.*, 2018).

For animals to benefit a lot from organic feeds, the feeds should be adjusted from time to time to ensure that an animal obtains all the necessary nutrients. This is because different organic feeds have different types of nutrients.

3.4 ARE THERE BENEFITS FEEDING LIVESTOCK ON ORGANIC FEEDS?

When it comes to organic Livestock feeds, there are numerous benefits that the animals and also the society at large could be harnessed.

One of the main benefits of feeding livestock in organically produced feeds is that the quality of animal products is improved. As for the livestock known to produce meat, with the help of organic feeds, they end up being healthier and strong hence the production of quality meat. For poultry farming, more so in the production of eggs, layers that have been reared on organic feeds produce eggs of high quality with high protein quantities compared to eggs produced by layers that have been reared on artificial poultry feeds. Unnecessary and unhealthy contents in animal products due to the artificial feeds should be eliminated completely.

The other advantage of the use of organic feeds is that they do not contain any additives. During the production of organic livestock feeds, no chemical additives are used, meaning the feeds grow naturally on their own without the boost from any artificial components. They only rely on Organic farm inputs, which in turn end up boosting the quality of the feeds and ensuring that the biodiversity is preserved. Organic feeds do not contain additives such as preservatives, for the packaged foods; flavourings and colourings ensuring that domestic livestock feed in quality feeds that promote their health. Basically, with organic livestock feeds, there is a guarantee that there are no harmful chemicals within the feed.

The third benefit of the use of organic livestock feeds is that there are no Genetically Modified Organisms. This is because organic feeds are purely fed to natural animals that do not require complex techniques during their rearing. Furthermore, genetically modified organisms are believed to be one of the main causes of untreatable illnesses among both animals and human beings in our current society. With a guarantee that the rearing of genetically modified organisms is not practised, consumers will have confidence in the purchase of different food products for there is the guarantee that the product is healthy.

The production of organic Livestock feeds ensures that no chemicals are used. The use of chemicals to destroy pests and diseases which damage organic feeds is discouraged and completely eliminated during the growth phase of organic livestock feeds. This ensures that there are no chemical residues left on the feeds for an animal to consume (Kongsted & Sørensen, 2017). The chemical residues are often left on the feed's surface or incorporated deep within the structure of an organic feed. The consumption of chemical residues does not only affect the livestock directly; but also affects the consumers of animal products indirectly by damaging their health.

The production of organic Livestock feeds is environment friendly. One, the chemicals used in boosting crop production are not used hence the chances of them being swept away into the nearby water sources are eliminated ensuring that the water sources are clean. The surrounding habitats are well preserved more so in term of their structure is not affected by the use of non-chemical farm inputs during the growth period of organic livestock feeds. The soil structure is improved for the use of organic farm inputs does boost soil fertility and structure in so many ways. For instance, organic manure not only contains important soil nutrients, but also contains soil microorganisms that boost the aeration of the soil among other benefits.

The use of organic livestock feeds reduces the cost incurred in the purchase of feeds for livestock. This is because organic feeds are produced locally, actually within a farm set up; making it easy for farmers to access the feeds and feed them on their livestock. Still, in terms of cost, the cost incurred in transporting purchased feeds from their production unit to the destined location is saved. This is because all a farmer has to do is manually move the feeds from the farm directly to the livestock, which is a short distance.

The use of organic livestock feeds does also save on the cost incurred in producing the feeds. For instance, artificially produced livestock feeds require the purchase of several chemical inputs, which require money for them to be purchased; however, with organic livestock feeds, the farm inputs are locally available within the farm structure. One, the manure produced by the livestock will be used in boosting the soil's fertility. Remains of organic feeds could be used as a cover crop hence helping in the conservation of soil moisture for the freshly grown organic feeds; suppression of weeds preventing them from being a nuisance to the newly grown organic feeds; boosting soil fertility for once the remains decay, they release important nutrients to the soil. Still, the remains of organic livestock feeds could be used as ingredients in the preparation of compost manure which is in turn used in the production of fresh animal feeds.

The production of animal feeds using organic methods improves the locomotion of animals. Organic livestock feeds are mostly grown on a large scale on a certain piece of land (Krieger *et al.*, 2017). Upon maturity, instead of a farmer having to harvest the feeds and give them to livestock; there is the option of letting the livestock feed freely on the feeds increasing the chances of them moving from one place to another as they feed hence boosting their health not only in terms of addition of important nutrients to the animals but also by allowing them to stretch their muscles.

Figure 3.3: An image showing one of the advantages of the use of Organic Livestock Feeds (The animals can move freely as they feed, hence stretching their muscles).

Source: https://ecolonomics.org/organic-livestock-nutrition-feeding-health-nutrition-of-organic-farms/

With organic livestock feeds, the chances of producing different types of organic feeds are improved. With this, a farmer does not have to incur too much cost on ensuring that their livestock feed on a balanced diet. This is because; different organic feeds could be grown on the same piece of land, but in different plots of land since different organic feeds require different methods of production. The production of different types of organic feeds within the same locality ensures that the livestock feed on a balanced diet from organically produced feeds.

The production of organic feeds for livestock guarantees food security. The climatic conditions are changing in almost all parts of the world. The use of artificial farm inputs is slowly degrading the soil quality and increasing the concentration of chemicals on animal feeds; forcing the animals to feed less on such feeds hence deterioration of animal health. With organic feeds, they are quite resistant to climate change and tend to easily adapt to new climatic conditions over time. Such characteristics prove that organic livestock feeds will defy all the odds and thrive ensuring food security in terms of livestock feeds.

Most organic feeds ensure the recycling of soil nutrients as they are being replanted within the same piece of land. This is because, a large percentage of organic feeds are rich in nutrients such as nitrogen, meaning they will restore nitrogen levels which are important in providing energy for the growth of the organic feeds.

Organic livestock feeds are rich in medicinal value. This is because they are one hundred percent natural and do not incorporate any artificial elements within their structure. Once an animal feeds on them is the same as feeding on herbal medicine. This improves the immunity making livestock resistant to disease-causing microorganisms. Still, most artificially produced livestock feeds are affected by pests and disease (Leeb *et al.*, 2014). However, with organic livestock feeds, their natural chemical structures make them resistant to disease-causing microorganisms and are less infested by pests.

Organic livestock feeds help a lot in dairy production. Most of the dairy cattle during parturition undergo serious health deterioration processes. One, they have to heal after giving birth; production of milk rich in nutrients for the survival of their newborn calf; among other processes which need energy. The use of animal feeds produced through non-organic farming is said to be not enough. This is where organic livestock feeds come into play. Organic feeds provide the dairy cattle with just enough amounts of nutrients to help regain their energy and also help in their phase of milk production.

The use of organic livestock feeds to feed domestic animals contributes to energy conservation by two ways first way that the energy used in the manufacture of artificial farm inputs such as fertilizers, pesticides and herbicides is saved for in the production of organic livestock feeds.

Second way is that with the production of livestock feeds organically, energy consumed during the distribution of artificial farm inputs from the point of manufacturing to the farm is saved. In the production of organic livestock feeds, mechanization is limited. This means that the energy used in powering machines such as field tractors is saved. With the production of livestock feeds organically, the rate at which the feeds are to be irrigated is minimized. This is because the organic farm inputs help the livestock feeds in developing deeper roots that draw water from the deep ground surfaces. With minimal irrigation, energy used in powering the water pumping system and supplying water to the irrigation channels is preserved.

Production of organic livestock feeds is preferred for its preservation of biodiversity. For instance, direct planting is practised during the plantation of livestock feeds ensuring that the soil is not interrupted in any way. Minimal soil interpretations boost the soil water retention mechanisms; improve soil structure among other advantages.

The use of organic manure in the production of organic livestock feeds boosts the soil structure. For instance, compost manure does contain microorganisms that play a big role in boosting the soil aeration; in the breaking down of decayed organic matter such as the uprooted weeds boosting the nutrient contents in the soil among other benefits.

Organic livestock feeds are produced in large quantities as compared to artificially produced livestock feeds. Bulk production of livestock feeds to ensure that the farm livestock has an adequate amount of food to last them for long, more so during the drought period. The rate at which organic livestock feeds are produced does also give hope to the farmers, more so farmers who major in livestock farming by assuring them of the continued availability of livestock feeds; which in turn gives hope to the world, for livestock do play a big role in the world through their by-products (Lu *et al.*, 2010).

The production of livestock feeds using organic methods does play a big role in reversing climate change and global warming effects or even at least trying to reduce the rate of climate change. The organic inputs used in the production of livestock feeds improve the soil structure and provide the feeds with enough nutrients for their growth. The nutrients in turn help the plants to develop deeper roots as they search for more water for their survival. The deep roots end up holding the soil particles together. Still, the soil microorganisms added into the soil through compost manure do help in holding the soil particles together. Soil aggregates that are well compounded next to each other reduce the rate at which carbon dioxide gas is released from the soil into the atmosphere. With a minimal percentage of carbon dioxide gas in the atmosphere, the rate at which the ozone layer is destroyed is minimized. This ends up slowing down or completely eliminating global warming hence no climate change. Still, the artificial farm inputs used in the production of non-organic livestock feed such as the ammonium nitrate fertilizer is known to release nitrous oxide gas which is a greenhouse gas and contributes to global warming effects. However, with the use of organic farm inputs, there is no release of carbon gas into the atmosphere.

Organic livestock feeds, which are of different types, are rotated from season to season to ensure that the soil nutrients are not depleted. The rotation boosts their resistance to pests and disease-causing microorganisms. This is because, microorganisms left in the soils after the harvesting of organic feeds, tend to be specific to a certain type of organic feed; meaning that, if the livestock feeds are rotated, a microorganism that caused illness to previously planted food type will not affect the new type planted. The same case does apply to pests for they tend specific to certain types of organic feeds.

Production of livestock feeds using organic methods of farming ensures that there is the local availability of animal feeds. With non-organic methods of production of livestock feeds, the availability is minimized. This is because the conventional animal feeds are globalized meaning that their demand is global. With global demand, the feeds might be fairly distributed but the probability of them reaching the most remote farms is almost zero. However, with organic livestock feeds, they can be produced in any part of the world hence they are locally available to all reducing the cost incurred in the purchase of the globally recognized artificial animal feeds.

Organic livestock feed as stated earlier is known to have high nutritional value. This is because, during their production, chemical fertilizers that incorporate nitrates within the organic feed structure are not used; lack of nitrates within the crop structure ensures that the plant nutrients are formed naturally and that the nutrients are in large quantities. The soil quality is high hence the trace natural trace minerals are available in high quantities; and are later incorporated into the feeds boosting their nutrient content (Lovarelli *et al.*, 2020). Lack of use of pesticides in the production of organic livestock feeds boosts their immunity for most of them fight pests and diseases naturally. Boosted crop immunity means that the crop has formed within itself enough nutrients to help it in overcoming basically any pest or disease attack.

Organic livestock feeds are available in so many varieties; unlike conventional livestock feeds which only rely on genetically modified seeds for their production. The conventional livestock feeds are strictly limited to the available manufacture seed varieties; but with organic feeds, different types are produced from the large pool of seeds available. The large pool of organic livestock feeds has been availed due to the different climatic conditions in different parts of the world; and also because of the natural way of saving seeds from each harvest for replantation purposes instead of having to purchase genetically modified seeds. Basically, the production of livestock feeds using organic methods promotes the practice of seed saving.

As stated earlier, organic production of livestock feeds employs minimal mechanization. This means that most of the workforce in the production of organic livestock feeds is manual and the manual work is for sure a lot; from the direct plantation of organic livestock feed seeds; preparation of compost manure; distribution of the compost manure in the farm; harvesting of the livestock feeds for them to be fed on domestic animals; among other roles. With the previously mentioned roles, the production of organic livestock feeds does provide different job opportunities. The job opportunities improve the living standards of individuals more so in-remote rural areas.

The production of organic livestock feeds promotes the preservation of soil from eroding factors such as soil erosion. As stated earlier, the organic manure used to boost soil fertility does also boosts the soil structure. Improved soil structure means that the soil aggregates are held together more tightly preventing them from erosion through wind erosion. The cover crops spread on the farm during the planting season do prevent splash soil erosion and also all forms of water erosion. This is because the cover crops cover the soil protecting it from direct conduct with the agents of soil erosion.

Production of livestock feeds using organic methods of farming presents a safe working environment. This is because the use of chemical farm inputs has been eliminated. For quite some time, individuals who practice conventional farming are exposed to the risk of different types of illnesses due to their continued exposure to chemical farm inputs. Artificial farm inputs contain some elements that pause a great danger to people, more individuals who have continually being exposed to such elements over and over again (Manuelian *et al.*, 2020). This exposure leads to illnesses such as cancer among others. It is true that with the employment of protective mechanisms, the chances of individual suffering are minimized; but so long as they are exposed to the chemical farm inputs for decades, they are still at risk of contracting dangerous diseases. With organic production of livestock feeds, farmers are not exposed to any kind of chemical elements hence the probability of them ending up being sick is zero.

Organic Livestock feeds are less contaminated as compared to conventional livestock feeds. As it has been stated over and over again in the above sections, the production of organic livestock feeds incorporates the use of purely organic feeds; whereas with conventional livestock feeds, artificial farm inputs are used. The artificial farm inputs always leave some traces in the feeds, and after harvesting, livestock ends up consuming the traces in the feeds that causes health complications to them. However, with Organic Livestock feeds, there are no chemical traces on them hence they are safe for consumption by livestock.

The production of livestock feeds through organic farming methods reduces the rates of leaking soil nutrients. The use of chemical components in the production of livestock feeds slowly changes the soil structure with time and they end up degrading the soil by making it weak. The soil is made weak by changing its structure such that it only relies on the nutrients obtained from the fertilizers for it to support plant growth. With the weakening of the soil, nutrients slowly leak to the lower layers of the soil. Continued leaking leads to contamination of groundwater table which in turn contaminated the rivers and lakes that rely on groundwater table as their source of water. However, all this is avoided through the use of organic farming methods. Organic farming methods instead of weakening the soil, make it stronger by adding naturally made nutrients that boosts the soil structure hence the soil can hold together its nutrients on the top parts of the soil; reducing the chances of nutrient leaking. However, mismanagement of organic farming techniques could also lead to nutrient leaking in the soil. For instance, when the same organic matter producing only a specific type of nutrients, and not variety, is spread on the farm over and over again, the soil becomes weak and the chances of nutrient leaking are increased.

As stated earlier, organic livestock feeds do contain several medicinal contents. This improves the immunity of the animal making it healthier. However, with conventional livestock feeds, the animals rely largely on manufactured drugs for treatment purposes. The continued use of manufactured drugs, does for sure treat the animals of their illness, but over time, the animals become resistant to the drugs. This is because, as the drugs are continually being used, the animal's immune system mutates and incorporates the drug structure deep into its system too; meaning that, once a drug is consumed, it will not have any effect on the animal for its structure and mechanisms of working has become part of the animal (Mench, 2008). However, the chances of the medicinal content in organic feeds becoming part of the animal's immune system are low. This is because livestock feed on a variety of organic feeds all of which contain different medicinal values.

Organic livestock feeds are more fresh compared to conventional livestock feeds. This is because most organic Livestock feeds are produced locally; hence they are fed to the animals after they have been freshly harvested from the farm. With conventional feeds, most of them were manufactured months ago by the time they are fed to the livestock. For conventional livestock feeds to last that long they must be preserved through the addition of preservatives to them; unlike organic livestock feeds which do not require any preservatives since they are fed on livestock directly from the farm. The freshness of organic livestock feeds boost the appetite of livestock improving their growth rate unlike the conventional livestock feeds whose added preservatives lower the appetite of livestock.

The use of organic livestock feeds has helped most nations to save up on the Agricultural incentives given to farmers by their governments. Such incentives have been mostly used to purchase artificial farm inputs such as the likes of fertilizers, pesticides and herbicides. With the production of livestock feeds using organic methods of farming, most governments no longer have to fund the purchase of fertilizers for free Organic manure is used; the purchase of pesticides is no longer required for the organic feeds contain medicinal components in them which can resist attacks by pests; the purchase of herbicides is no longer needed for cover crops spread on the farm can suppress the growth of weeds hence the organic feed does not have to fight for plant nutrients, sunlight and other elements required for plant growth (Mingmongkolchai & Panbangred, 2018). The save incentives money could be used elsewhere; for instance, it could be used to find the development of infrastructure, betterment of the education system among other sectors.

As stated earlier, organic livestock feeds are produced locally hence available locally to the most remote farmers; which is a practice that promotes unity within a community. With conventional livestock feeds, most farmers had to purchase the feeds via brokers who would make a deal on their behalf with the main producers. However, since organic Livestock feeds are produced within a community; the farmers can sell the surplus feeds directly to other livestock farmers within their community. The direct selling and buying deals promote the unity of the Community by building good relations within the community members. Still, the demand for organic livestock feeds is slowly growing. This is because many individuals believe that organic feeds will boost the health of their livestock, improve the quality and quantity of the animal produce among other beliefs. With the increased demand for organic Livestock feeds, farmers can make more income which improves their living standards.

Apart from the use of organic livestock feeds as food to domestic animals, they could also be used as bedding material for the animals. This is because most of the feeds produced using Organic methods of agriculture are quite soft hence good bedding materials. This might sound like wastage to some individuals, but it is not the freshly harvested Organic Livestock Feeds that are used as beddings, but the remains after the livestock feed on them. The remains are spread in bed-like structures in cattle sheds and used as bedding materials. This saves up on the cost incurred in the purchase of cow mattresses.

In terms of marketing, organic livestock feeds amass very high profits to the farmers. This is because the production costs are low and the selling price is high due to the increased demand for organic livestock feeds globally. Basically, the cost incurred in the production of the organic livestock feeds is so low because; one, there is no purchase of fertilizers for the Organic methods used to increase the soil fertility in long term periods; two, the control of pests and diseases is done organically without the need to purchase artificially manufactured pesticides; three, the control of weeds is not needed for cover crops help in suppressing the weeds preventing them from growing; four, there is minimal water requirements for the Organic methods of farming improve the soil water retention capability. The production of Organic Livestock feeds only incurred costs in terms of funding the labour force and maybe during the harvest season; so, the comparison between the production costs and the profits made is like saying that, Organic Livestock feeds are free to produce.

3.5 CHALLENGES AND SOLUTIONS IN PRODUCTION OF ORGANIC LIVESTOCK FEEDS

The production of organic livestock feeds is faced with few challenges some of which may be easily solved.

One of the challenges is that the requirements needed for a farmer to be certified to produce organic livestock feeds are quite a challenge. This is because not all farmers are quite well educated on how to meet such requirements since most of them are in remote areas. However, this challenge could be solved by ensuring that several workshops are conducted, in all regions, even the most remote regions; educating farmers on methods of producing livestock feeds organically, and the easiest way with which they could meet the certification requirements.

The second challenge is to do with labour. As stated earlier, the production of livestock feeds using organic methods of farming is quite labour-intensive; from the preparation of compost manure to its distribution in the farm to the spreading of cover crops and harvesting. The issue of labour could be solved by, innovation of mechanization methods which will promote little interruption of soil structure; for with organic farming, mechanization is discouraged because the current techniques lead to soil disruption.

Production of organic livestock feeds is mostly associated with tradition, for most of the organic farming practices have been copied from our ancestors. This makes it to be more centred among the rural communities and it discourages investors to invest in the production of organic livestock feeds for they have almost no clue in what they are investing in. Such a challenge is being solved through availing of information on the production of organic livestock feeds on all social media platforms so that investors can understand what goes on. As more and more investors learn about organic livestock feeds and their way of production; they are encouraged to invest in the production of Organic Livestock Feeds making it a success, not only in rural areas but across the globe (Mie *et al.*, 2017).

In conclusion, the future of the entire agricultural sector lies in organic farming. The future of livestock production lies in organic livestock rearing. The future of the healthy, quality and quantity production of livestock feeds lies in organic livestock feeds. May the world embrace organic farming; may the world embrace Organic Livestock rearing; may the world embrace Organic Livestock Feeds; for that is where the survival of the entire Agricultural sector for ages to come lies, that is where the survival of both the livestock and the human race lies.

Chapter 4

Limitations of Organic Animal Production

CONTENTS

4.1 INTRODUCTION

Organic livestock production involves the production of a huge number of food products where many rules apply to maintain better food quality. The rules demanded are normally on the welfare of animals while at the same time considering the care to the environment, production of animal products that are healthy for human consumption and limited use of medicine. The International Federation of Organic Agricultural Movement is one sector that is responsible for ensuring that all the rules concerning organic livestock production are followed. The sector has been responsible for coming up with or rather writing down the intentions on livestock produced organically. As far as most farmers are concerned, the rules with regard to ensuring that healthy animal products are produced are not clear. In fact, when the sector was drafting the rules to be followed, consequences for not keeping organic livestock healthy were not stated. In organic livestock production, a farmer should always have a strategy to ensure that there is disease prevention when it comes to the livestock they are handling. As a matter of fact, when it comes to disease prevention, the rule is that the animal should be seen to exhibit behaviour that is considered natural. On the other hand, the animal should be fed, but the food should be organic and also have the ability to cope with infectious diseases, unlike animals that have been raised conventionally.

This shows that there will be a very limited chance for the animals to get sick but if this is the case, the farmer should know of various alternative methods to treat his animals rather than use the conventional ways. In this case, the producer of organic livestock should be well conversant with several natural remedies that are used to treat various animal diseases or find a vet that deals with herbal remedies. According to the already stated regulations, farmers are normally advised on the use of homoeopathy and phototherapy. However, not most farmers can use this method because of the lack of evidence on the effectiveness of using such methods. In most cases, an animal raised organically has the likelihood of contracting an infectious disease when they are grazing in the field. This is so because, while in the field, the animals are exposed to bacterial, viral and parasitic infections. Exposure to the outside environment has been one of the major causes of conventional livestock being unhealthy. Such diseases caught while in the fields are the major reasons for animal products to be unfit for human consumption. As such, producers of organic livestock are normally urged to try having preventive measures to ensure that such occurrences are avoided (Miranda-De La Lama *et al.*, 2017).

They are rather urged to first keep better breeds of animals, maintain hygiene in the area in which the animal sleep, addition of acid to the water that is drunk by the animals and perform a pre- and pro-biotic. When the area is known for instances of infectious diseases affecting their animals, the urge to farmers is that they need to understand such diseases and ensure that they administer vaccines to their animals.

There are various characteristics of organic livestock production that make the venturing by farmers into such a business, very complicated or in other cases very difficult. In the current society, most members of society are trying to move towards the consumption of organic products. This is attributed to the health issues that are experienced in the world. As the world is becoming industrialized, the environment is becoming more and more dangerous to humans. Agricultural industrialization has led to the use of modern technology to influence the natural growth of living things. Take for instance beef production. Research by scientist's claims that in several years, cows may become extinct especially with the high rate of consumption of beef products. As such, some scientists have been able to devise ways in which beef can be made in a laboratory. This entails slicing a small piece of animal meat from a surface that can regenerate easily. After the piece has been sliced, it is then injected with some enzymes that will ensure the growth of the product enough to feed a whole family. These methods are seen to be effective especially when considering saving cows from extinction. The problem with such a method is that it will mean that the beef will be synthetically produced which in most cases means, unfit for human consumption. This among many other reasons has led to consumers demanding more organic livestock products. It is known that organic livestock production ensures that there is better treatment and welfare of animals, the products do not contain unwanted chemicals and at the same time, the environment is kept safe. Ideally, the European Union has taken up the mantle to provide regulations in the production of organic livestock products. This has been done for the sole purpose of ensuring that consumers do not have confusion about the brands of different organically produced products. In the EU, all member states are required to comply with such rules to ensure that the best is provided to the consumers of the products.

In many ways, organic livestock differs from conventional production. When the system of production is organic, the space that is taken up by the animals a normally larger compared to the space taken by conventional livestock. Apart from this, the animals are fed organic products and there is always straw bedding. In the case where there is a shortage of organic feeds, farmers are allowed to feed their animals a certain percentage of organic feeds to compensate farmers in this case are forbidden from using antibiotics and in case any treatment of the animals has taken place, the period it takes to use the products is normally longer (Nalubwama *et al.*, 2011). These measures ensure that the animals are reared under welfare and the products remain to be quality. With all these facts, different organic farmers have spun up, all of them with different ideas regarding how to rear organic livestock. If the various states are concerned with the production of organic livestock products, then all the ideas by farmers regarding ways of rearing such breeds of animals should be taken into account. When the EU was coming up with various regulations regarding organic livestock production, there were not clearly stating the consequences of not following the rules. This has become quite a limitation especially when it comes to its production. When consequences are not clearly stipulated, the person that ends up on the losing end is normally the consumer.

Organic livestock farming is a sector that scientists and agriculturalists are trying their best to revive.it goes without question that the introduction of this section will be fruitful both health-wise and has the potential of earning income to those who practice the same. As a new thing, consequently, there needs to be learning and direction to be successful. Only a limited number of farmers know the same even though it seems like it is going to be part of our future. As a great factor in society, this sub-sector in agriculture has some limitations. Some of the limitations can be easily dealt with while some are quite complicated. Some of the limitations are provided below.

4.2 ANIMAL WELFARE AND HEALTH PROBLEMS

Researchers claim that the main problem when it comes to animal welfare and diseases is knowledge transfer. It is only in few circumstances that that new technical research would be required when dealing with animal welfare and health. Such circumstances include, when dealing with animal feed and when dealing with organic feed. The main problem that normally faces organic livestock farmers is when dealing with bad hygiene standards. This occurs in almost all animal species that are categorized as organic livestock. Such diseases may include mastitis, ketosis that is a metabolic disease, lameness, which normally affects dairy cattle among others. There are some reasons why this situation is still in a worse state.

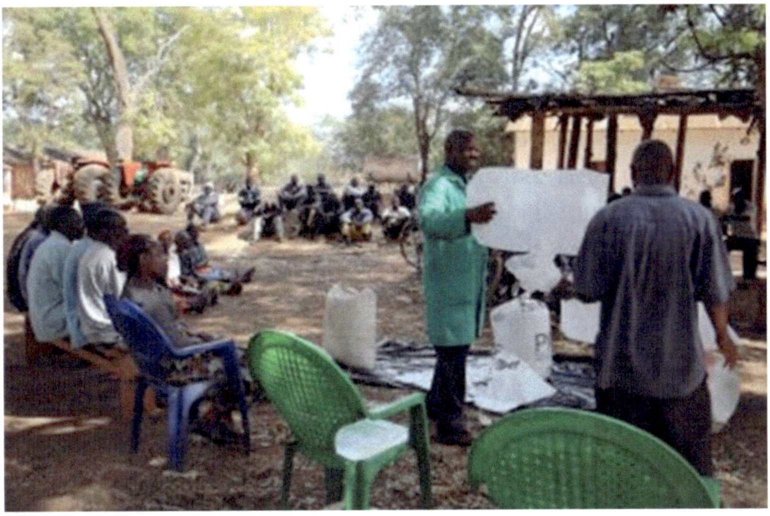

Figure 4.1: Education is key in ensuring that organic farmers have knowledge regarding livestock keeping.

Source: https://www.ifpri.org/project/developing-local-extension-capacity-dlec

One main reason is that it is very difficult to find organic livestock farmers who only specialize in the same. As consumers continue demanding high-quality products, farmers must accept to specialize. In any case possible, to produce high-quality products, the producer must put their hearts minds and souls in the same. For example, a mango producer. If the producer is putting all of his time and attention into the production of pineapple, they normally experience high yields of the same. Take for instance a farmer who cultivates a variety of crops and on the other side is a dairy farmer. In such circumstances, each sector yield in the farm can not normally have the ability to reach its full potential. In one way or the other, comparing the yields in the farm with that of another producer who is investing the same amounts, one would find out that the yields of the one specializing are normally high. Livestock normally requires attention compared to crops. The producer should always be present to observe the changes that may occur to the animal. When an infectious disease affects the livestock or one of the livestock, the producer should be able to observe such cases to ensure that they do not infect the rest of the livestock (Papatsiros, 2011).

Another reason is that the advisory system that deals with organic livestock is not well specialized. When it comes to consultants, it is quite easy to find those that are all-around but may prove difficult to find those that have specialized in the same. It proves to be hard to find those that have knowledge of the fact, which makes dealing with these situations hard as seen above, livestock requires a lot of attention and in addition, requires one to know the same. Organic livestock farming is a new concept in society, which means that only a few individuals know how to deal with circumstances affecting them. As a matter of fact, if an individual visits a specific region, finding individuals know the same is normally slim to none.

The farms in which organic livestock are normally raised tend to undergo huge economic pressures. This especially happens to pig farms and dairy farms. The huge economic pressure normally occurs because of the poor prices of organic milk and pork and sometimes eggs. When one goes to the store, milk that has been placed in packets are normally sold at higher prices compared to milk that is sold in containers. Since organic products are new in the market, most of the owners of the stores are not willing to buy the merchandise to sell in their stores. Individuals fear change in society, most specially business people. Individuals that work at the store will always want to have the merchandise that their customers are used to. The problem comes in when the consumers buy products from the store because of the loyalty they have towards a certain brand or distributor. This makes the lives of organic farmers quite difficult because of the lower prices they have to sell their products. It is well known that organic feeds from the market are normally expensive and thus with low prices, one may only be able to reach breakeven. This means that it may prove to be quite difficult for the producer to make any form of revenue from rearing organic livestock.

The last problem occurs because of wanting to ensure that one practices proper organic animal husbandry. Some components such as the bioaerosols and mycotoxins are found on the straw that may lead to various health complications by the animal. With an organic livestock system, there are restrictions on the use of preventive treatments. The livestock normally has contact with the environment and the soil, which is not free from microorganisms, which may form as the root cause of some of the deadliest diseases in livestock. When it comes to horses, farmers that are breeding organically have been banned from using anti-parasitic disinfection in the horse stables. These are just some of the animal husbandry procedures that an organic producer is needed to follow. They are supposed to ensure that the products from the livestock remain to be quality (Pérez *et al*., 2019). However, as seen, they tend to have consequences on the animals themselves. As such, organic producers should always be on the alert to ensure that none of the above precautions taken affects the livestock.

When it comes to organic livestock products, individuals everywhere demand that there be quality in whatever they produce. The quality should not only be on the product but also on the process undertaken to ensure that the product reaches the final consumer. The main task that is left to the organic producers is for them to seek the required knowledge regarding how to rare organic breeds of livestock and what needs to be done to ensure that the products are of high quality and that customer satisfaction is achieved. The poor economic returns from organic livestock farming are one of the major problems that are currently being looked into since it leads to the demotivation and destruction of efforts of even the most hardworking farmer. it is with no doubt that organic farming should be treated with uttermost importance since most of the health problems currently facing human beings can be solved by the use of better management practices, especially on edible food and nutrition.

4.3 LEGISLATION AND ANIMAL WELFARE IN ORGANIC SYSTEMS

To ensure disease prevention when dealing with organic livestock, the assumption or rather the belief in the mind of individuals is that housing, feeding and care of the animals are normally based on the animal's natural ability to combat diseases. In case of the occurrence of a disease, some regulations have been put in place especially by a government state to ensure that the animals and their products remain to be of quality. These regulations may prove to be of great assistance to the farmer and the environment, though in most cases, may prove to be limiting. This is so because some of the regulations that have been put in place may prove to be expensive to implement or in other cases may even prove to be harmful to the livestock as discussed. Earlier, it was discussed that animals need homoeopathic and phytotherapeutic measures to cure a certain infection in livestock or when trying to prevent the spread of a certain kind of disease. These measures are not that familiar to most of the breeders of organic livestock, which may prove to be quite challenging. Phototherapy involves the use of plants and plant extracts to generate a herbal remedy that may cure a certain disease that has infected livestock.

Homoeopathic products on the other hand are culinary products that are extracted from plants, animals or even mineral substances. As one can see, these products are natural remedies and may prove to have little or no side effects on the livestock. The problem however is that majority of individuals in the current society have no clue on how to handle homoeopathic and phototherapeutic products. The use of the above products is one regulation that has been set by the EU in the case where an animal has been infected b an infectious disease. In fact, the regulation claims that in case of infection by a disease, the organic livestock is to be treated by the use of phototherapeutic, homeopathic and trace elements and products instead of chemically produced products. Such chemically produced components include chemically synthesized allopathic veterinary medicine and antibiotics. The plant-animal and mineral components can however only be used when their therapeutic effects have been proven to be effective by the scientists. This is mostly for the case whereby the elements used are appropriate for the condition in which it was intended to work on (Prunier *et al*., 2014).

In other circumstances, the use of such remedies may prove to be ineffective. In such a case some regulations have been put in place to ensure that proper measures have been put into account for the betterment of the animal and the maintenance of the quality of the products. The EU has a regulation that covers the same.

Figure 4.2: The rules by the EU requires that organic livestock are treated using homeopathic and phytotherapeutic ways.

Source: https://organic-plus.net/author/organicplusnet/

This regulation claims that in the case where the above products do not comply with combating the disease it was intended to, it should be made sure that the animal is put in a position in which it will not be under stress or any kind of distress. In such a case where treatment is very necessary to avoid the same, the farmer is then allowed to use chemically synthesized allopathic veterinary medicinal products. The farmer can also apply the use of antibiotics but this should be under the responsibility of the veterinarian.

In the interpretation of the above-stated rules by the EU, the inclination that an organic farmer should have is that the use of homoeopathic and phytotherapeutic products should always be the ones in mind before they resort to the use of chemical products. Conventional treatment in this can only be used when the natural remedies have not proven to be successful. When it comes to correct interpretation of the EU regulations, then individuals should understand that the use of conventional medicine when dealing with organic livestock is allowed if and only if, the natural remedies do not affect the disease. This is whereby the conventional remedies are described as alternative methods of treating the disease. However, there is insufficient data concerning the EU regulations stating the effectiveness of the use of a particular remedy on a certain species and on the particular disease intended to be cured. The regulations are also not clear whether the homoeopathic and phototherapeutic methods of treating a disease that is currently used by the farmers of organic livestock comply with the regulations provided by the EU. In most countries, most stakeholders are trying to find out whether the various remedies that are used should be recognized by the state or need to be withdrawn to ensure that the animals receive the best treatment instead of speculations that may hurt their health and quality of the products. When one analyses the medicine that is used to treat the animals, it is clear that the majority of the organic farmers prefer the use of alternative methods of treating their animals rather than the use of natural remedies. This is attributed to the fact that the majority of the farmers lack evidence that shows the effectiveness of the natural remedies.

The use of pharmaceuticals and antibiotics for preventive purposes is normally not allowed among farmers of organic livestock. The use of such remedies is when the animal is sick and there are no available remedies that can be used to cure the same. Growth and production stimulants such as coccidiostats and hormones are not allowed either when dealing with organic livestock. In fact, the use of conventional methods to treat animals is only allowed several times in a year. When dealing with organic livestock, the withdrawal periods of animals on a particular treatment remedy are normally longer as compared to conventional livestock. Furthermore, the waiting time after the use of conventional medicine is also longer. Take, for instance, when conventional animals are treated for a particular disease, say mastitis, it will take close to one week before the producer starts to milk the cow (Röös *et al*., 2018). However, when the animal in question is one reared in an organic farm, it may take close to two months for the farmer to be able to consume the milk. This may prove to be quite expensive or rather inefficient, but the assurance is that the product that will be ingested by an individual will be good for their health, as they will have no traces of the medicine used.

Some regulations regarding the management of organic livestock have not yet been made clear. With these facts, it is with no doubt that, the organic farming business is being put into jeopardy while the regulations by importers are set so high that it is becoming hard for the local producers to expand their markets worldwide. this is a huge limitation to the organic livestock farming sector as they cannot be able to access markets in other parts of the world it is no doubt that there exists a niche worldwide about the supply and sale of organic products. This shows that these products have the potential to improve the economy of a particular country. as such, to promote this business, authorities need to be advised on coming up with regulations to make other countries confident on their product hence open up the international markets for the particular product. Ideally, most countries and with no doubt the EU are currently trying to revise or rather update their regulations concerning organic livestock farming.

4.4 MARKETING OF ORGANIC PRODUCE HAS NOT BEEN PROPERLY STREAMLINED

As a consumer, food safety always comes first. Individuals are normally scared of food safety. The human body is quite fragile and thus caution is normally taken when choosing the products to consume. In most cases, individuals are normally scared of animal products. This is so because; there are rumours of livestock that are ready to be consumed within six weeks. Looking at such animals, they are normally healthy, full of weight and are seen as fully-grown animals. This is not even close to being remotely possible. In such circumstances, animals are fed with food supplements and injected with growth hormones to make them appealing to the human eye. This is the reason why most countries put strict regulations when it comes to the importation of products. It is with no doubt that, the stakeholders of a country would always want to protect its citizens and ensure that they always receive quality products.

In 1995, Hassall and associates had various comments about the marketing of organic livestock products. According to them, there are products from organic livestock that can be traded or rather provide an opportunity to the producers of the particular product. Such products include yoghurt, whole milk, butter and cheese from cows and dairy products from sheep and goats. For sheep, their wool is also in high demand. As a matter of fact, overseas, wool is highly demanded. The Australian organic wool is one product that is in high demand and countries from Western Europe and Japan are normally placing enquiries on the same. If such interest in this particular product could be taken seriously, then this can be converted or rather state as a product with great sales potential. There is however a greater limitation when it comes to this product. It is at farmer or rather producers lack a scouring facility that can help processors in the scouring of the wool and further increase the quantities available to sell to the customers of the same (Sorge *et al.*, 2019).

In a local setting, the demand for organically produced products is normally on the rise, especially among affluent and health-conscious individuals. It may be quite easy for an individual to have faith in products that are produced locally rather than those that are imported. This is especially for the case of meat and other animal products. An opportunity in the line of animal products is the certified lamb and beef. Lamb is normally soft and sweet meat and is normally in high demand. In fact, those that are organically produced are always in demand. With the case of pork and poultry products, the demand is normally quite low because there are available feed rains that have been certified.

There are several factors that affect the marketing of livestock products, especially meat and meat products. Requirement for certified slaughtering and processing facilities is one of the most significant factors that are affecting the marketing of livestock products. This might prove not to be a big issue since, in every business, there needs to be a certificate or rather a permit that allows an individual to continue with their business. With this, continuity of supply is normally an issue with the producers. It is with no doubt that, the rearers of organic livestock are limited and thus it is quite hard for the supply to be able to meet the always-rising demand. In such a case, the producers are put in a position in which they need to pull their resources which in turn leads to inconsistencies especially when providing quality products and those with better grades. For some producers, it may prove to be quite easy to sell their products. An example of such organization is the Riverina Organic Farmers Organization, which cooperatively markets and sells their product. In this case, this is a cooperative society that is well known worldwide which makes it easier for them to sell or rather market their products. However, for smaller or rater local smallholders, it may prove quite difficult for them to market their products on their own. Such producers need the help of their government to access international markets, which at times is a problem because of the regulations put in place by the international markets.

Limitations of organic livestock farming can be brought about by problems relating to marketing. Some of the marketing problems include that which involves quality consistency. This is normally brought about by the food that the animals are fed, the genes of the animals or the farm practices by the farmer. Inability to ensure continuity of supply is another marketing problem, which tends to tarnish the integrity of a product. To ensure that there is an improvement of product inconsistency, some of the producers have decided to devise a strategy. The strategy involves them designating many finishing products or properties that are located near the processor by developing a certain genetic base that is common (Sonoda *et al*., 2018). The diverse geographic distribution of producers can overcome the continuity of supply. In this case, the producers will have integrity in their supply and there will be minimized substitution. In this case, the producers package the products in small quantities and avoid large quantities and then they package them and brand them to be sold in a retail market or the food shops.

End markets for products can be classified in many ways. They include direct delivery, wholesalers, the foodservice industry, retailers and the export market that is normally eyed by almost all distributors. The price for the product when it is being sold in the market is called the price premium and is normally identified in the market. The climatic and environmental conditions also have impacts on the supply of organic produce. Take for instance drought. Pasture is the main source of food for organic livestock. When there is drought, most pasture dries out which leaves the livestock hungry and forcing the producer to buy supplements for the livestock. Since the producers of organic livestock normally wish to maintain the quality of their products, they buy supplementary feed that was only grown organically. This sometimes is quite hard to find and the producers pay a huge price to acquire such products. Since they are hard to find, producers have to undergo additional costs to purchase the products. In this case, most of the producers are faced with very tough choices of whether to sell their livestock or rather feed them conventional feeds. If they do so, their stock of livestock will lose their organic status. Even if the livestock are recognized as organic and are fed non-organic products, they cannot afford to be sold as organic in the market as they will have lost their organic status. However, they are still suitable for use as the organic farmer can breed other organic animals. Their offspring can then be sold as organic.

This shows that in seasons in which the environment is faced by drought most of the organic farmers tend to de-stock their livestock. In the case whereby the farmer loses the core breeding stock, the farmer can be very devastated. In the case, whereby the farmer is breeding to resist the internal and external parasites is considered to be very important when looking for proper animal husbandry. In this case, the farmers are not allowed to use synthetic pesticides to control the pests. Various organizations have taken up the role of careful selection and breeding of livestock that has a resistance to pests and diseases in their program. When there are interruptions to the market of the product, may it be due to a hundred-year drought or the seasonal livestock growth patterns, some of the processors tend to outsource their supply, which leads to the producer losing their market (Ankamah-Yeboah *et al.*, 2019).

In other cases, which are normally the worst-case scenario, non-organic products are normally sold in the market as organic products and are labelled. To ensure that there is consistency in supply and quality of the products from the producer to ensure that they have market alliances or rather business associates that will ensure that their markets can run smoothly whenever they are faced with a bumpy situation. When there are production alliances in the organic business, those who need access to the products can easily access them through a single avenue while supply is assumed to be stable since the suppliers are many. In such a case, the suppliers do not worry about being dumped along the way for missing a supply.

The major problem, however, is that most of the organic producers are agriculturalists which makes them typically poor in business. They tend to come up with poor strategies to work their part in the case of the distribution of their products. An effective supply chain is considered very important in ensuring that they have successful marketing whether they are acting individually or in a group. In the case where the farmers are acting as a group or in one, they should ensure that all the members of the group comply with the rules and regulations set by the group. Another thing is that the group members should be chosen in terms of whether they have an interest in the entire marketing process. Those alliances that consist of several producers always do a follow-up of their products to ensure that their products have smoothly gone through the entire value chain process. In most cases, they ensure that the product has been able to reach the final consumer such that they can be able to know the brands that are preferred by the consumer or rather the complaints put forth by the consumers. For a legible producer, there are some factors that they have to look a before deciding whether to distribute their products through a supply chain.

Organic credibility is one alliance that needs to be addressed. In his case, the producers need to understand and most importantly implement the organic certification to ensure that the products are allowed for selling by the proper authorities. As such, the producer needs to have certification showing their compliance with the law. The second is that they should ensure that no matter what, there should be an all-year-round supply no matter the various challenges that they ace on the farms. They should also ensure that the products maintain the same quality all the time. To deal with cases whereby the products may not be able to match the demand, the supplier should be able to outsource the products to ensure that the customers remain happy. Another thing that a smart supplier or producer does is that they are careful to hire a processor who can be sympathetic to their aims especially when meeting demand becomes quite a challenge because supply is limited (Wilhelmsson *et al.*, 2019). The processor selected should understand the roots of organic distribution and production and the requirements of organic farming. In this case, the producers will be forced to be innovative and on the other hand, should be willing to allow their processors to take part in the entire supply chain. Apart from choosing a processor that is sympathetic to their aims, they also need to choose appropriate trading partners. In all businesses, all stakeholders that take part in a certain business should be able to understand each other and work together to prevent misunderstandings in the group.

Marketing is a great limitation to organic livestock production but can be handled well if all the rules of marketing are followed to the latter and business partners are chosen with great care.

4.5 DECLINES IN ORGANIC PRODUCTION IN THE FIRST YEARS OF PRODUCTION

During the first years of production, organic livestock farmers tend to go through losses and it is important for organizations and even countries to offer premium prices that may help the producers to continue with their reduction. Such prices will act as incentives to the farmers for them to continue with their production and produce high-quality products. As reviewed above, maintaining an organic livestock farm may prove to be quite expensive and with poor prices being offered some of them will force to dig in their pockets or even threaten to leave the business. Organic feeds are not as cheap as they may seem as a matter of fact; organic feeds are more expensive when compared to conventional feeds. As such, keeping up with farmers who practise conventional farming may prove to be quite difficult. Furthermore, in the first few years when the organic farmer is still starting their business, they can easily lose a number of their initial stock. This is so because it would be the first time the farmer is trying out something new and quite delicate. As such, the animals require proper and expertise handling and thus any mistake made may lead to losing some stocks. This may lead to other producers of the same to opt-out of the business and those that make it use many funds to acquire resources.

The only advice is that; proper extension services should be provided to potential organic livestock producers for them to survive in the sector. In fact, the government should hire extension workers to educate members of the public on the same.

Figure 4.3: Organic livestock feeds are quite expensive compared to conventional feeds.

Source: https://www.indiamart.com/proddetail/organic-animal-feed-4218034097.html

4.6 OTHER LIMITATIONS

Internal and external parasites may be a problem for farmers who are practice organic livestock keeping. This is because such parasites are preventable by the use of synthetic medicines. However, organic livestock should not be vaccinated by the use of chemically synthesized medicines. As such, they can be easily be attacked by such parasites which may lead to the loss of the livestock. These are huge limitations and therefore the research society should be under pressure to ensure that they come up with herbal remedies with preventive measures against such parasites.

Breeding is one other thing that is a limitation to the organic livestock community. When it comes to conventional livestock, genetics involves the high production of their species. In the case of an organic farmer, a breeding animal is chosen for the various qualities it has such as resistance to diseases and the ability to survive in harsh conditions. The problem is that, if the animal has been genetically modified in a way, it is not going to be permitted under organic livestock. Artificial insemination is also not allowed when it comes to organic livestock production (Åkerfeldt *et al.*, 2020).

In conclusion, organic livestock farming is the future of the world and needs to be practised by several individuals. Practice should be done and education on the same should be provided to ensure that those new in the industry do not completely fail in their first attempts.

Chapter

5

Organic Pigs

CONTENTS

The rearing of organic pigs is a growing economy all over the world as it has been giving exotic prices to farmers practising it alongside providing numerous environmental benefits and superb welfare for animals. Organic farming has been around for more than 80 years but it is a niche that has been slowly creeping into the mainstream market. The view of consumers has switched to organic food and organic certification bodies are present to oversee that food getting into the market has all the organic production qualities.

5.1 A GUIDE ON THE CERTIFICATION OF ORGANIC PIGS

Pigs can convert grains, unsaleable agricultural products into high-quality pork which is marketable all over the world. Its waste products can also be used as high-grade manure. There are federal guidelines that have been outlined in place to certify pigs as organic. This ranges from the feeds they are given, the health they undergo, animal source and the living conditions they are exposed to. Organic agriculture is the practice of producing food and fiber that utilizes and optimizes the use of natural resources surrounding the area such as soil by enhancing its use. Ecological processes and cycles are utilized rather than using chemicals to help in production on the farm. Third-party agencies work hand in hand with organic farmers to ensure that all the guidelines are being followed that will ensure that pigs are organic certified. An organic system plan (OSP) is formulated.

In the United States of America, under the National Organic Program (NOP), every certified organic farm must have an organic system plan formulated. The organic system plan is an outline that clearly explains how you plan to use your farm to deem it organic by showing steps and activities that you the farm will be partaking in terms of the feed being given to animals, the healthcare programs, the source of animals in the farm and the living conditions the animals are being exposed to. The organic system plan should satisfy all the requirements given outlined by the National Organic program. In the United States of America, pig farming that has been organically certified takes a very small percentage. In as much as there is a high demand for organic pork, the growth is small because of the premium prices being charged in the market. The high prices are a result of the high cost of maintenance and production of organic pork which forced the prices to up for the farmers to foot their expenses and at the same time get a profit margin (Von *et al.*,2004).

On farms that practice mixed crop-livestock farming, organic pigs can play a very critical role in being recyclers. There are organic grains that are deemed not consumable by humans and therefore such grains can be fed to pigs, which in turn convert the grain into high-quality organic pork. Also, there are products that cannot be economically stored and sold to consumers therefore they are fed to organic pigs and give you high-quality organic pork. Organic production in farms prohibit the use of chemicals in the production process and this means that a farmer should find other means that are organic to help in the production process that are nutrients and supplements for crops. Pig manure can be an excellent source of nutrients for the crops. Four major aspects have to be looked at in organic pig production to deem it certified organic. These factors are

- Source of animals.
- The feed given to the pig.
- Health administered to the pigs.
- Living conditions being exposed to the pigs.

5.2 SOURCE OF ANIMALS

The most common way in which farmers start to raise pigs is by getting feeder pigs and taking care of them until they reach a marketable weight. However, for a pig to be sold as organic, it has to have undergone all the required guidelines stated by the USDA. These guidelines should have been exercised at least thirty-eight days before a piglet is born. This means that a pig will be deemed organic only if the requirements have been practised at the onset of its third trimester. The sow will not be considered organic but the piglets will. A non-organic sow or gilt that has been bought from a farm can produce an organic piglet provided that the guidelines to follow to achieve an organic piglet have been followed. Even though a farm has been organically managed for years but the pig is not furrowed using an organically managed dam or water source, then it cannot be sold as organic meat in the market. It, therefore, does not qualify as an USDA-certified organic slaughter stock (Alonso *et al.*, 2020). Organic cull sows have a very thin market due to the difficulties undergone when taking care of organic stock. Most of the pigs are sold in the conventional market with the current conventional prices.

5.3 PRINCIPLES OF ORGANIC PRODUCTION.

Below are some of the principles that should be considered when breeding organic pigs;

- You should employ renewable resources in your farm and agricultural activity.
- In every stage of production, the farmer should be able to maintain the organic integrity and important qualities of the organic products by emphasizing that everything is left organic at every process throughout the production.
- The health of the pig and other livestock should be a priority and therefore one should employ alternative care to promote this.
- In every stage of production, you should reuse and recycle materials to increase effectiveness.
- The biological diversity in the farm should be maintained at all stages of production.
- The long-term fertility of the soil should be adhered to by employing organic processes in maintaining this.
- Care about the environment by protecting it at all costs by minimizing soil degradation, Soil erosion, avoiding environmental pollution, optimization of biological productivity and enhancing the overall Ecosystem within your jurisdiction.

5.4 RAISING ORGANIC PIGS

Pigs are normally kept in flocks and can either be put in barns or an open area. They are commonly grazers and therefore they can access food at ground level or low level. For pigs that are put in barns, they are normally fed at least twice a day but most of them are let to just graze around the field as they look for grass and natural vegetation. Pigs need water and there provided for in troughs and containers although in areas that have a lot of moisture the amount of due in the grass they eat is enough to sustain them through the day. In countries such as New Zealand, most of the time there is enough due in the grass to sustain them most times. Once the piglets have become of age, they become weaned from the mother and are fed on hay, pasture and grains. This process is vital because of the increased competition between the piglet and the sow for food. Moderate temperatures are conducive for the survival of pigs. Do not be fooled by the thick bacon it processes. Just like humans and other animals, wind chill and heat index factors play a major role in determining the optimum conditions of a pig. The atmospheric temperature can be different from the body's temperature. During the cold season that is winter, it is vital to check on the cold stress in pigs (Vanhonacker *et al.*, 2010). One should check on symptoms such as shivering, teeth grinding, huddling, and reduced grazing period times which subsequently increases food intake in the barns. Appropriate facility management practices should be implemented such as; Installation of windbreaks, necessary adjustment of inflow and outflow of air, dry bedding and putting up shelters.

Installation of any heat source equipment in their shelter is also important especially for vulnerable livestock like newly born piglets, freshly shorn pigs, a piglet with reduced body temperatures and ill livestock. Farmers are advised not to use heaters equipment that is used for poultry (brooder piglet). Rather they should get heat source equipment conducive for pigs. The piglets should be placed on high hanging on ceilings to avoid the occurrence of accidents. A shelter will require to be well protected against harsh environmental conditions such as wild predators which might take advantage of vulnerable piglets.

Shelters should have adequate space whether it is temporary or permanent. This is because there should be room for expansion. Fully grown ewes require space of about 13 to 17 square feet each whereas pigs should have space of about 16 to 20 square feet each. For the pairs of dam and offspring, an additional 14 to 20 square feet should be provided. Space during feeding should also be considered. This is because large animals might harm or injure small animals when feeding. If there is simultaneous feeding of both large and small animals, then 13 to 21 inches should be allowed for each pig. Vertical bars should have a width of about six inches.

A constant and plentiful supply of water should be the number one priority in the shelters, especially during the dry season. In the case of a lactating pig, an average of about 6 gallons should be provided. During the cold season, water should be heated to keep the animals at their required body temperature and to avoid dehydration since they become reluctant to consume water. Placing water troughs along the outside walls should be avoided during cold seasons as it may enhance the freezing of the water. Doors of the shelters should be located in a direction where there will be maximum sun exposure and electricity installation on shelters might be vital to enhance warmth in the shelters. It also helps with enabling proper airflow. The shelter should also be placed at a higher ground to avoid the possibility of flooding.

Adequate ventilation is important in shelters as it enhances the health of the animals. Even during the cold season, there should be proper air ventilation to eliminate dust, gases that are emitted from the decay of manure, excessive moisture and pathogens that might be lingering in the air. This is because the combination of these products can cause lung damages, bronchitis and overall loss and reduction in yield. Few factors can indicate that there is poor ventilation in the shelters which include heavy condensation on walls or ceiling, strong manure odour, animals with runny noses and/or coughing and excessive heat emanating from a shelter. Installation of curtain walls and cupolas and fans can help facilitate ventilation. Wood is the best material to use in constructing shelter. A properly pressure-heated wood will eliminate the possibility of animals chewing on them. Wood is considered to be the best construction material because it provides conditions that can help in controlling temperature.Concrete is the best for flooring although it can be costly and might get cold during the cold season thus it should be adequately bedded (Bessei, 2018).

5.5 NICHE OF THE PIG MARKET

There is a large market all over the world looking for organic products from pigs ranging from meat, sausage, bacon Etc. Organic products are already booming in unions such as the European Union and African Union. From 2015 to 2016 the demand for organic pig products increased by 5 pc. Countries all over the world have more than enough consumer demand for organic pig products. In Canada, organic pig production has a very small specialized market. This is because there limiting factors that need to be addressed sooner or later. One of the limiting factors is that in Alberta there is a shortage of certified organic feeds and most pig products are thus not considered organic leaving available organic products to be limited in supply and very expensive to obtain. Another limiting factor is that for piglets to be deemed organic the piglet meat needs to be handled in an organically certified and regionally inspected meat processing facility. With such constraints, it makes it very expensive to set up a business that deals with organic pigs as that means that supplements need to be transported from other areas and products cannot be found locally.

In conclusion to set up an organic pig business in Canada, one needs to have a solid business plan and one needs to ask these questions listed below:

- What and where is your target market for your organic pig products? How large is the network? What type of pork cuts does your target market need? If you cut the piglet are you able to sell all of it at once and what are your customers willing to pay for your piglet?

- What is your competition? Are there any other businesses providing the same services? What share of the target market do they have control in? What is the period they have been operating their business? How has their business been thriving?
- What is the required quality acceptable in the market? What is the right procedure to follow in every phase of production I.e. production, processing, transportation, marketing and selling? And will your target audience pay enough to sustain your needs?
- Is there a certified slaughterhouse within your area? Where is the location of the organic piglet processor? Are they willing to cater to your demand and supply needs? How many piglets can they take? When do they need the piglets? What kind of intelligence does the manager of the meat processor able to follow and record? Can the intelligence gather be given to authorities and customers when needed to? What are the costs?
- Are there legit and well-certified organic materials in your location? How readily available are the organic feeds in your area and is there a contract that needs to be signed? What kinds of feeds are available? Do they match my needs in terms of quality and quantity? Are they seasonal? Is delivery of the feeds readily available throughout the year or is it available only during the harvest period? Is the quality of feeds testable by relevant bodies? What will be the expenses incurred i.e. buying of feeds, trucking, handling and storing?

5.6 ORGANIC PIG PRODUCTION

As shown above, every jurisdiction must comply with the organic production standard that has been set. Just because someone has pigs and rears them with organic pasture and vegetation and does not give them any chemical or artificially manufactured drugs does not mean that the product is deemed to be organic. Strict rules and regulations in all the production phases must be adhered to in order for a piglet to be considered organic. Organic piglet standard of production must be able to have followed the forage-based system which involves the way you graze your pig, management of pasture on which the pig live and the provision of the appropriate non-pasture feeds over the non-grazing winter period. Organic pig production systems need to develop depending on the location where it is situated as there are factors such as climate and feeds which vary in every area. Every organic pig production must have financial viability for it to function and that is why a feasibility test should be undertaken before venturing into such a business. Every farmer is also required to be responsible for his or her organic piglet meaning there are no trade-offs (Buller *et al.*, 2018).

To have a successful piglet management system, there are areas one needs to put focus on at least one year prior before venturing into such a business.

- **Market-** one should be able to identify and understand the market to which they will be selling their organic piglet products that are the identification of the target market, the criterion to which the market expects you to deliver your product and the nature of the contract agreement with the market.
- **Flock nutrition-** the farmer should be able to understand the quality and specification of the feeds that the pig will be taking to optimize the quality of product, access to nutritional advice regarding your feed is also vital and finally food testing and ration balancing.
- **piglet health and well-being-** kind of care, preventive strategies that will be employed to prevent diseases, approved medication, regular check-up and advice from the veterinary clinic.
- **Cost of production management-** one should focus on these main factors that are labour and animal feed, nurturing and developing skills necessary to manage organic pig production and looking for someone who can advise you on matters regarding financing your business.
- **Facilities and efficient handling systems-** a farmer should create a framework that will optimize operations by enhancing efficiency in every production phase and the model should take into account both human and animal welfare for example piglet feeding.
- **Fall back plan-** one should be able to account for unforeseen circumstances such as drought, floods and diseases.

Figure 5.1: Most countries require at least 365 days period of planning to certify that a pig has an organic standard.

Source: https://agricultureandfood.co.uk/news/news/2019/may/20-special-regulations-ensure-even-higher-welfare-standards

5.7 CRITICAL ACTIVITIES

For your piglet to be certified as organic, there are vital activities that must be practised. They involve;

- The feed is given to the pig should be a flock-certified organic feed.
- The standards and requirements needed to manage pasture for the pig should be organic.
- Any artificial substances or substances that are deemed not to be organic should not be exposed to the piglets in any way.
- Slaughterhouses should have organic standards and need to be certified

Clients or consumers require their quality organic products to be planned, recorded and maintained in terms of the following components.

- In terms of rations of the organic piglet pieces, they are required to be at least thirty percent forage at any particular time.
- Minerals and salts must have an organic certification.
- Apart from winter, pigs should be allowed to eat pasture.

- In the final phase which is processing and handling, it is required that all activities should have organic standards.
- The distributors of straws for bedding should ensure that it is organically certified.
- The wood used in barns and structures should not be treated with preservatives and they should be certified organic.
- During transportation, certified organic material (animals, feeds, bedding) will at no time be in contact with non-organic certified materials.
- Any vehicle transporting non-certified organic materials is also required to be thoroughly cleaned in between loads.
- Any party given the duty to either transport or handle certified organic materials must sign a declaration that his or her equipment is up to standard.

5.8 KEEPING CLOSE TABS ON THE ONGOING BUSINESS OPERATIONS

A piglet business will need a well laid out plan with good records of every financial and business transaction. In the case of the organic piglet business, a more detailed and meticulous laid out record-keeping is very vital. It is important to look for certifying agencies that are strict with their guidelines as it has been witnessed in most places that sub-standards certifications are being issued which is bad for business. All information needed should be collected and recorded accurately and clearly. After recording, good storage of the records will come in handy and finally, the stored information needs to be audited by relevant authorities (Broom *et al.*, 2013).

A certifying agency needs the following information to be collected to see if standards have been met.

- A field history of at least three years regarding your acre land involved with all the organic production activities.
- Acreage maps.
- A recording describing the rotation of crops and pastures.
- The identity of all piglets is deemed to be organically produced and all their particulars like health records.
- Records showing all land inputs beginning from farming tools, supplements, feeds and labourers attending to the organic pig.

- Records of methods used in making manure and compost materials
- Products and procedures employed to control pests, weeds and diseases.
- The systems of storage are used to store farm tools and inputs.
- The description and type of input used.

In Canada, under the Canadian pig identification program (CSIP), a radio frequency identification tag should be installed in all piglets before they leave the land where they were born. In the process of organic piglet and pig production, these identification tags will be vital during record keeping. For identification, these tags are normally placed on the right ear whereas the other tags are placed on the left.

5.9 APPLICATION FOR ORGANIC CERTIFICATION

The process of certifying crops and livestock is quite lengthy and it might take more than three years. There must be at least 3 years from the last time any agricultural substance was applied any prohibited substance to make it have an organic standard. From the well-recorded details regarding your organic transition processes, there will be an audit to certify your worth.

Some supporting documents that might be needed include the record of the purchase of feeds, breeding stock, tests conducted on soil and water. Record of tools used during the organic production period will also be required. All purchased supplies over the period should also be given. Once the certifying agencies have approved the certificate, then the farmer can go ahead and make other investments.

5.10 CONVERTING A CONVENTIONAL PIG TO AN ORGANIC PIG

Numerous advanced calculations are employed in the process of converting a conventional pig to an organic pig. Before the onset of the third trimester, a non-organic sow can give birth to offspring that are organic provided a transition is done before then. For a sow to be given the organic status it should have the following characteristics;

- The sow should have unique individual pig identification.
- Have completed relevant records such as identification, breeding and dates.

- Switch to using organic materials such as bedding, feeds and supplements.
- Records of the conversion process of the program such as the date, feed, storage and capacity
- The sow should not at any time been exposed to any contaminated substance during the transition.
- Should have records of a continuous three-year organic management program on pasture and forage.

Piglets that were born from non-organic pigs can have the organic status provided that they have maintained the organic management program from birth.

Figure 5.2: Newborn piglets from conventional sows can be deemed organic.

Source: https://www.thepigsite.com/articles/simple-guide-to-organic-pig-production

5.11 BUILDING A BUSINESS

5.11.1 Managing of Cost

The price of organic materials in grocery stores looks eye-catching as they a mostly affordable in many places and from a business perspective that means the returns will be high. Operations in the farm significantly change when one decides to shift to organic farming from commercial farming. A huge limitation is the strict rules and guidelines which will eventually reduce the number of pigs meaning lower returns in organic pig farming. There is also the limited supply of organic feeds which might be a constraint. This might force the farmer to venture into growing the organic crops by themselves to cater to the demands and in turn make income for themselves. Other costs might be incurred because there will be the need to do cropping meaning additional farm equipment needs to be bought and a subsequent introduction of a new management technique (Brito *et al.*, 2020). The introduction of cropping might alter the economies of scale that might have been the sole source of minimizing costs.

With the volatile nature of the pig market shifting to organic pig farming might be a challenge as such farmers rely on large-scale selling of conventional pig within a year. In the event of the existence of a secure organic market, organic pig farming might come in handy as it would reduce the volatility in the market. Animal feed takes the largest chunk of the product resulting in up to 70 percent. Organic feeds and supplements might just push that cost higher. The sale of bacon might be used to increase income but it is known that the sale of pigs is the largest revenue producer.

The second-highest cost of production is labour. Every stage in production requires a certain form of labour ranging from breeding, weaning, weighing, health care, shearing, grading and management of predation. Therefore, the farm manager needs to come up with a model that will optimize the operation and at the same time lead to effective use of resources. Time is also seen as labour because the farm will need to hire professionals and they get paid by the hour.

Other costs that might be incurred include security costs, construction costs, energy costs, processing costs, distribution costs, etc. A study was conducted in France to see the price premium needed to help offset the extra costs incurred due to the transition from conventional to organic pig farming. The study came up with a few assumptions.

- Organic pig bred over an extensive system required a higher price premium as compared to conventional pig production.
- Feeds and labourers amounts to upto 70 percent of the production cost in a conventional pig business.
- An intensive and hastened production system is not conducive for organic pig production.

- In a conventional system, the intensive process is powered by feeding grains to pig which is prohibited in organic pig production.
- Irregular cash flow will always be experienced in pig production while the overhead cost will always be constant therefore one should seek financial advice regarding setting up organic pig production.

5.11.2 Market Opportunities

In Canada the per capita pig consumption has increased by more than 10 percent since the early 2000s. However, the consumption market is thin. Only 2 percent of the Canadian population takes organic food and most of the 2 percent take organic vegetables based on the perception of quality, health and sustainable food production. Even though organic meat is purchased by a small percentage of the population, there has been a rise in the demand for that small niche. Certified organic pig producers can take advantage of that market opportunity. According to research, only 21 percent of meat consumers in Canada take organic piglet and this can rise if the quality keeps improving and the consumer pricing of the product gets adjusted. The study showed that retail pricing is the most important factor. It shows that pork has a higher price compared to other commonly consumed types of meat in the market.

A good organ pig producer should look for his target market that are willing to pay a premium on the piglet meat. Consumers choose organic piglets as their choice of meat because they want to avoid livestock that is fed from chemically sprayed crops and others don't want animals fed from grains. Such consumers understand they will need to pay for more but there is a limit to what they are willing to pay (Cambra-López *et al.*, 2010) This will have to force organic farmers to be creative in finding a way of selling their products at a premium that will enable them to keep their businesses afloat.

5.11.3 Planned Marketing

This is the biggest obstacle facing organic pig farmers because they need to find a viable market willing to consistently buy their product at a price that will cater to expenses and contribute to a margin. If the farmer has access to a large market, direct marketing can be the best move to use. With little knowledge and experience in the marketing of organic piglets, the farmer needs to set up his own price and look for his market. The farmer needs to ask himself about some marketing plans he can employ.

- Who will purchase my product?
- Will I have enough customer bases to cater to my supply?
- Are my customers will pay more for my products?
- How can I expand my customer base?
- What part of my products would my customers love to buy for tailoring purposes?
- Is the different taste of organic piglet meat appealing to my customers?
- Can my premium charge be able to cater to my expenses?
- Are there certified organic pig processors in the region?
- Will the packaging of my organic meat be appealing to customers?
- Is the processor willing to store frozen piglets?
- How much will it cost to process piglets?
- What will I incur in the process of handling, delivering and institutional payments?
- Advertising costs?

5.12 PRODUCTION MANAGEMENT

5.12.1 Breeds and Breeding

- There are large numbers of different pig breeds all over and there is necessarily no good breed than the other. In the case of organic piglet production, cross-breeding offers the best quality and qualities of different breeds is crossed

Ideally, it is difficult to find two different breeds of an organic pig to cross-breed therefore conventional pigs are cross-bred and the best quality breed is then the transition process of converting it to become organic begins. When doing replacement breeding, conventional pigs can be used though they cannot be resold as organic pigs if they have been owned for less than a year. That is why it is highly emphasized that accurate record keeping be maintained to avoid any sort of confusion. Unlike getting organic pigs for cross-breeding is hard, there are numerous places you can get pigs with quality, different genetics. It is recommended to visit different areas to look for the most desirable genes. Biosecurity should be upheld by wearing protective gear such as boots and overalls during the visit. It is advisable to choose all your pig for a specific look to avoid the contact and transmission of diseases. In the real world, there is nothing like a disease-free pig but you can find the well-maintained pig. If you need pig from replacement breeding you should go to designated farms and them from there rather than going to a feeder pen or a slaughterhouse. It is also important to check on flock dispersal so that you can be able to located pigs of the same age group to avoid compromising health security. Avoid auctions as flock dispersal is significant. It is obvious that all pigs were first raised from pasture but you will find that other genes do better in the grass system as compared to others.

Organic piglet producers emphasize various types of traits. Quality maternal traits imply that there will be a subsequent quality economic trait. Such maternal traits involve:

- Ability to mother
- Longevity and toughness.
- Thriving well under organically managed systems.
- Capabilities of foraging, produce sausage and maintain body condition on pasture
- Good health is increased resistance to diseases.

One should also consider buying a pig with a specific trait depending on your preference. You should be able to have in mind what your ewe flock needs to produce. Either you want piglets with quality maternal traits or quality slaughter piglet traits. This of course is dependent on the kind of target market you are aiming for (Castellini *et al.*, 2012). Your flocks might be producing either bacon or sausage meaning you might need to diversity your product horizon. Look for a pig that you think will give desirable traits to the offspring. In most cases, birth vigour that is the piglet survival is the most preferable option. A Growth rate is also preferred. Pigs that will bear offsprings that are easy to herd on pasture will eventually give you good returns as they are less expensive to maintain.

Get information from the person selling you the pig? Ask for the best quality pig and be ready to pay extra. Besides, you will be buying quality. Market piglets are best produced by terminal sires as they have good genetics for growth and carcass. Maternal breed sires will give you offspring that have quality products.

Artificial insemination (AI) is an acceptable practice in organic pig production. It is a bit more expensive but it has the benefit of preventing diseases from coming into your farm from other animals, therefore, maintaining a high quality of your organic pig.

There is the prohibition of the use of estrus synchronization and embryo transfer. Since hear detection is impossible with pigs practising artificial insemination without performing synchronization of the ewes heat cycles might lead to reduced rates and increased cost per piglet being born. A pig that can survive after the grazing period is long overdue is quality for the organic pig system and the combination of genetics and good management leads to pig resilience.

5.12.2 Feeding and Nutrition

In Canada forage plays an important role in feeding organic pigs. It is grown and harvested to be used after the grazing season. With an average of 200 days during winter, farmers need to plant forage and store them once harvested. One might face challenges during winter as there might be inconsistencies in supply which might cause starvation of organic pigs. If found it will be more expensive than planting yours which doesn't improve efficiency in your farm business operation. In order to get a constant supply of organic feed you will need to build a rapport with your supplier and that might be disadvantageous to small-scale farmers as they may not fulfil their end of the contract. Delivery and storage will also add costs. Certifying agencies play an important role in helping you find the best source of feeds and supplements tailored for your organic piglet. Take advice from them.

You should sample feeds and perform a test on them to find out quality feeds tailored for your organic pig. With this, you will be able to know the feed nutritional value and make the proper rations for your pig. Organizations that sell supplements might help you with that. Feeds should have a certification that they are organic. Though there can be exceptions in the event of extreme conditions. Genetically modified organisms (GMO) are prohibited as they are not compatible with the principles of organic production. Supplements should be grown using non-genetically modified organisms (canola, corn and soybeans). Sausage from organically certified livestock should be used to feed piglets.

Organic pigs that have been fed solely by grass tend to grow slowly as compared to grain-fed pigs and this can change their taste and quality in terms of tenderness and moisture. Therefore the farmer should always look at a market base that will appreciate such kind of meat (Cornish *et al.*, 2016).

5.12.3 Grazing

To get organic certification, records should show that flocks are extensively grazed on pasture and forage. Pigs being ruminants, can easily convert plants into meat, bacon and sausage. Some studies have been conducted and have connected extensive grazing of an organic pig with profitability. Nursing piglets are usually grazed hand in hand with their mothers until they are weaned. After weaning they are then transferred to paddocks that have a lower quality pasture. It is critical that the nutritional value of pasture is up to standard for the ewes and the young piglets to ensure proper growth and thus tests must be conducted. In addition, good quality pastures means that organic pigs can be sustained during winter and enable them to reach the desired market weight, reduce losses due to illness and demise.

Due to the short farming season in an area like Canada, having quality pasture will enable the farmer to offset the cost incurred from grazing the land. The grazing field should be tested and regularly checked on fertility, organic matter and water. There are techniques employed to extend the grazing season and they include forage stockpiling, swath and bale grazing and rotations between annual and long-term pasture.

5.12.4 Predation

Organic producers should put security systems in place to cope with the threat of predators that people on your piglets. Reports have shown that coyotes cause a lot of trouble in the area of Alberta. Guard dogs should also be used.

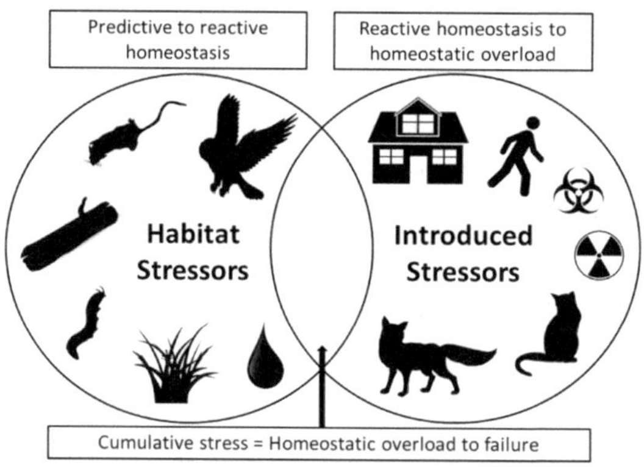

Figure 5.3: Combination of security systems to cope with predation.

Source: https://peerj.com/articles/9104/

5.13 HEALTH OF FLOCK

In organic farms, health is a great priority and minimizing illnesses is their concern. The general well-being of the organic pig should be looked upon.

Below are some points that farmers should adhere to have healthy flocks.

- Minimize the number of brought in animals.
- Establish tight biosecurity protocols.
- Create a management system that reduces stress and care about the welfare of pig.
- Make sure that feeds have the required nutritional needs in their rations.
- Expectant sows should be given special treatment and the newborn piglets should be fed with colostrum.

Medication is prohibited and if used, clear records should be noted down. The use of vaccines is also prohibited but in some special cases, it can be allowed for example in the control of clostridia bacteria diseases. Vaccines that are given in the right manner will not harm the carcass of the organic pig.

5.14 PARASITE MANAGEMENT IN PIG

The organic ship is heavily related to worms. Over the recent years, studies have been carried out regarding the persistent menace of worms noticed in organic pigs and it is a problem that needed to be researched. In the year 2008, a Canadian scholar and grower came up with a practical handbook called 'LIVING WITH WORMS IN ORGANIC PIGS'. This was written by a Canadian organic farmer called Peter Stockdale. During the same year, areas of Canada experienced the wettest seasons with a 40 pc higher precipitation rate than normal. It was discovered that wet conditions increased resistance to eradicating worms. It was realized that wet conditions increased anthelmintic resistance which added a whole other perspective to the challenges faced in parasitic control.

To achieve parasite management, research in Canada was conducted on two organic pig farmers. One by the name of Atcham had a flock of 150 ewes and 20 does but have been reduced to 60 ewes and 8 does participate in the pig parasite research conducted by the University of Guelph in partnership with the organic agricultural centre for Canada. The research found out that his farm had very low internal parasite counts as compared to other farms. The other participant is Peter Smith who is a certified organic pig and vegetable farmer who currently has 25 organic pigs with the aim of increasing them to 50. Smith does not use grains but purely pasture to feed his pig and uses hay combined with vegetable cells once in a while (Clonan *et al.*, 2015). In the research, none of the course facilitators has a background in organic pig farming and therefore they aim to collect knowledge. A convention was held in February of 2008 and 23 of the participants were pig and goat farmers but most of them were mainly pig farmers who mainly dealt with organic pigs but others had conventional pigs.

From the research, there were main points that were highlighted. The challenge of parasite control both affected organic and conventional pigs. However, according to Paul and Achim, it was found out that the use of organic soil, pasture and other feed management strategies there were numerous merits as compared to conventional farming when it came to parasite control in pigs. It was concluded that organic. pig practices significantly improved the environment necessary to help control parasite management in pigs. It was also found that it is convenient that the flocks should be pastured in paddocks and should be frequently moved from one to another to avoid the growth of parasites. This should be done especially in wet weather conditions as they facilitate and provide conducive conditions for parasites to thrive. Once the parasites start forming in the intestines, they lay eggs and once excreted from the animal they are left in the paddocks. The eggs then hatch, then the parasites climb on top of the grass and they are consumed by the pig which they will go back to the intestine which will lead to it being overwhelmed. Depending on the soil, weather and climatic conditions, the livestock should not be returned to the same paddock for weeks or even months to enable the parasites to die due to starvation. Not only will the parasites die due to starvation, but they will also be. Killed by the heat of the sun and also the pasture around the area will grow back and any manure would have decomposed by then. The overall parasite load around the paddock would have been reduced because its life cycle will have been altered.

The health of the soil is also very important, which in turn leads to the overall health improvement of the organic pig. The health of the soil could be improved by increasing nutrients, which is done things pasturing and adding manure to the soil. Regular soil checks should be conducted to see the level at which the soil is thriving thus determining if organic amendments should be added to the area. Paddocks should be diversified with different varieties of vegetation such as hay, grass and cloves. This is to benefit both the soil and the livestock. Some weeds should be introduced to the paddock to have a benefit with the livestock, as they may provide anti-parasitic effects when ingested by the pig. Weeds such as chicory and wormwood help draw up micronutrients from the subsoil up to the roots of the pasture. During the harvesting of hay, it should not be cut to a bottom level as it washes away the nutrients from the topsoil.

It was also found out that grain feeding can be difficult for a pig to digest and might alter its pH and in the long run, grain feeding can affect the overall health of the pig, which might subsequently lead to more health complications. Breeding is another factor that might impact parasites in pigs. Peter selectively breeds his pig to enable his piglets to become robust. Achim culls his pig but lets them freely breed. Culling is mostly done on pig that portrays undesirable traits or that have chronic conditions. With both techniques, there has been evidence to show there is increased resistance to worm pressure.

When it comes to the housing of pigs, factors such as exposure to sunlight, cold and ventilation were considered and it was found that prolonged stay in shelters might increase the chances of pigs getting parasites. In the shelters, sanitation, proper ventilation and thick bedding are critical. Proper flock management and soil quality improvement generally led to the improvement of the health of the pig. Even with the presence of parasites, they were not sufficient to lead to its death and the pig lived perfectly (Clark *et al.*, 2016). Several natural dewormers have been permitted to be used in organic pigs and they include garlic tonic, Hoeggers herbal dewormer and diatomaceous earth to mention but a few.

Chapter 6

Organic Poultry

CONTENTS

6.1 INTRODUCTION

Organic poultry is the raising of poultry from day-old chicks till maturity without the use of synthetic pesticides, antibiotics, hormones and mammalian by-products in the poultry's feed. Inorganic and GMO feed ingredients are not allowed at any cost. Organic poultry has recently risen to popularity due to people's decision of having food from natural sources without chemical ingredients. Many people have a mindset that organically reared poultry products such as broiler chickens are safer and far more nutritious than those grown under traditional conditions.

6.2 ORGANIC POULTRY PRODUCTION

With an increasing demand for organic poultry products day to day, organic poultry production has been the most sought-after activity. This has therefore created a need for education in aspects involving organic poultry in both producers and consumers.

In organic poultry production, there are certain conditions that are very important factors to obtain organic certification.

Organic poultry is produced under organic farming. Organic farming is a form of agriculture that involves integrated and environmentally sustainable production systems to give out acceptable levels of crop and livestock nutrition. Organic poultry and agriculture complement each other for example the crops need manure which they can acquire from the birds while birds on the other hand need other necessities such as bedding which can be provided by the crops (Chander *et al.*, 2011).

6.3 BROODING OF ORGANIC POULTRY

Brooding of poultry is the rearing and handling of chicks from the first day when they are hatched to about eight weeks old while providing them with necessities for growth such as heat and definitive care.

The chicks are placed in brooding units which are designed to house chicks during the brooding process and provide the required supplementary heat. Once the chicks have brooded, they are transferred from the brooding units to growing pens and they are only transferred if the broilers need to be sold or the pullets are needed to be transferred to permanent laying houses when they are about twenty weeks old. Once the pullets start laying, they are moved into pens or cages until they are slaughtered and sold out when the laying period ends which is around seventy-eight weeks.

Brooding in a Natural Way

Natural brooding occurs when chicks are raised naturally on the farms with the help of hen mothers and only a few of them are raised within a year. This is so because the number depends on the hen brooding ability and is usually about fifteen to twenty chickens.

Supplementary heat to the chicks is provided by the hen and therefore the hen should be in good health to prevent contagious infection to the chicks.

Artificial Brooding

Artificial brooding is the management of day-old chicks and raising them without the help of the mother hens. It introduces the use of a foster mother to the chicks who are temperature-controlled. Artificial brooding has been preferred to over natural brooding over time because;

- There is no limit to the number of chicks reared per year at they can be reared all round the year.
- There is control over the sanitary conditions therefore lesser cases of disease infections.
- The temperature may be regulated because the foster mother is already temperature controlled.

Feeding may be undertaken according to plan and can be easily managed since all thousand chicks brooded can be easily handled by a single person.

Figure 6.1: Brooding house.

Source: https://www.livestocking.net.

A brooding house should be good and dependable for regulating temperature conditions, steady flow of fresh air, enhance dryness, enough lighting, enough space, easily disinfected, give protection from chick predators, protection from unexpected fires, and eventually, a building should be cost-effective.

Care for Young Birds in the Brooder

Brooder and chicks inside should be managed effectively to ensure the chicks inside grow well to maturity and the brooding house remains in good condition over a long period.

To accomplish this, one should:

- Adjust the temperature of the room as required by the chicks and according to the accredited temperature of the organic council. In cases where oil is used in heating ensure the stove and lamps are in a good condition to prevent any sudden occurrence of fire incidences. Place the lamp in an area that cannot be accessed by the chicks (Celi *et al.*, 2017).
- Supply clean and fresh water to the birds at the very least two times a day. This is why before establishing the farm there should be a reliable source of water around.
- Avoid a damp poultry house as this may lead to a rise in the rate of certain infections to the chicks. Instead, use a deep litter system.
- Keep a steady flow of fresh air in the house regularly.
- Discourage litter eating by the chicks to avoid disease infections.
- Often review the fittings, temperature controls, feeds, and water trough structure before shifting the chicks into the brooder. Before doing something, the house should still be in decent shape.
- Always have a veterinarian within reach for emergencies . The veterinarian should be available at all times and have a manageable price.
- Check in on the chicks every morning and inspect their conditions and check their faces for any sign of abnormality.
- The brooder should be a certain level above the ground to prevent running rainwater from getting in and should also be at a place where the wind does not get in.

- The brooder house should have a specified number of chicks occupying it to prevent overcrowding. Every chick should have enough space to move and run about as this facilitates fast growth and immortality.
- Once a vaccination schedule is established, it should be followed regularly.
- Poultry older than 3 weeks old should be given chopped green grasses as it is beneficial and acts as a vitamin A supplement.
- The brooders and feed hoppers should be cleaned daily.

Equipment Necessary for Successful Brooding

Brooding is an effective management method in poultry processing, so knowing the appropriate brooding equipment is essential. This is significant because brooding aims to raise healthy chickens by providing them with the proper temperature (that of the mother hen) until they can adjust their body heat to the ambient temperature.

For successful brooding, some equipment must be provided: they include:

- ***Brooder Guard:*** A brooder guard is a kind of hardboard that can easily be made into a ring-like or circular shape. It is used to contain the chicks within a particular heated space. In normal cases, poultry farmers place 100 chicks within the brooder guard which is one meter in diameter. A brooder guard is not necessary where battery cages are used for brooding.
- ***Brooder or Heater:*** There are several types of brooders and heaters on the market. The four most famous are as follows:
- ***Charcoal / Kerosene Stove:*** This type of brooder is most applicable in areas where there is no electricity. An improvised charcoal or kerosene stove uses charcoal or kerosene to generate additional heat to warm the chicks.
- ***Hover:*** The most popular kind of brooder is this one. It has a 250-watt bulb—typically five well-arranged luminous bulbs or infrared bulbs with a power on/off button—in its heating unit (D'Silva *et al.*, 2017). To deflect the induced heat back to the surface, the hover is usually covered with a circular or angular metal layer. It's normally suspended from the ceiling by a cable or rope, and it can be adjusted to any height level based on the desired temperature.

- ***Electric Brooder:*** This is a thermostatically operated heater that evenly distributes heat over a large region. This type of radiator is beneficial because it stops chicks from crowding directly under the heater or brooder.
- ***Gas Brooder:*** unlike the electrical brooder, this type of brooder uses gas to generate heat.
- ***Infrared Bulb Heater:*** The heat needed by the chicks to keep their bodies warm is also provided by an infrared bulb heater.
- ***Automatic Heater:*** A 600-watt heating element is enclosed in an automatic heater, which also has a thermostat and a reflector. It's hung in the same way that a hovercraft is.
- ***Thermometer:*** Within the brooding house or pen, a room thermometer is needed. It's used to keep track of the brooding pen's ambient temperature. The room thermometer tells you whether the temperature in the brooding pen is too hot, too cool, or too cold.

How to Prepare for Brooding

Inspect the brooding house properly for any cracks or leaks two weeks before the poultry arrives. Repair the house if broken and thoroughly clean, disinfect and fumigate the inside of the house with an appropriate disinfectant preferably quaternary ammonia and leave the house to rest.

- Two or three days before arrival, cover the floor with suitable litter material about 7.5cm deep. On top of this, place rough thick paper.
- Place the brooder box in the middle and then arrange feeding and watering equipment radial from the centre of the hover. The feeder should be half under the hover and half outside. All this equipments are then enclosed by a brooder ring 45cm high. This guard can be made from strong cardboard paper.
- Check all heating and lighting equipment to ensure that they are working properly.
- Two hours before brooding starts, the feed should be placed in a feed-trough. The drinkers, which should have been thoroughly washed, are filled with clean cool water.

- When the poultry arrives, they should be quickly removed from the boxes and examined individually for defects and rapidly placed under the brooder.

What to do when brooding broiler chicks or poultices

- Once the chicks are placed in the nest box, they must be closely monitored to ensure that they receive enough heat and are not exposed to draught. The chicks' behavioural responses will tell you if the ambient heat is sufficient or whether there is a draft.
- When chicks or poultices crowd hover or near the heat source there is insufficient heat, where they move farther from the heat source, there is too much heat; when they spread out evenly there is sufficient heat. When there is a draught, the chicks huddle to one side.

Temperature Requirement for Broilers

Day	Temperature
1-4	32.3-38 °C
5-14	29.4-26.7 °C
21-28	26.7-21.2 °C
28-56	18.3 °C

Floor Space Requirement for Broilers

Age	Floor space cm^2
Week 1	100-120 cm^2 per chick
Week 2-4	250-300 cm^2 per chick
Week 5-8	700-800 cm^2 per chick

These show that there is a need to expand the space as the chicks increase in size.

Adequate ventilation: Ventilation of the brooder pen is not allowed for the first two weeks. During the first four days of brooding, the shallow feeders are used to supply feed to the chicks. An example of shallow feeders is a thick sheet of paper or egg trays. Clean water is supplied every day.

Other than taking care of the chicks by controlling temperature, ventilation and providing feed, the following daily routine operations must be followed from morning to evening in the same order.

Remove the dead chicks from the group to prevent those alive from feeding on them and potential infection by disease-causing organisms. Fresh feed is provided and constantly activated by hand.

The water troughs are removed and washed properly before they are filled with fresh clean water.

Some special operations such as de-beaking, deworming and vaccination are stress-inducing in the short run but are very advantageous in the long run.

Feeds and feeding during brooding

Broilers

The feeding system for broilers is as follows;

- Day 1 – Week 4 = Broiler Starter
- Week 5 – Week 8 and above = Broiler Finisher

Cockerel

- The first Day 1 – Week 6 = Chick mash
- Week 7 – slaughter = Grower mash

Layer / Pullet

- Day 1 – Week 6 = Chick mash
- Week 7 – Week 17 = Grower mash
- Week 18 and above = Layer mash

6.4 REQUIREMENTS FOR ORGANIC POULTRY PRODUCTION.

One of the basic requirements before starting an organic poultry farm is properly managed and suitable vegetation in the area.

Organic poultry producers need to practice ways of preventing livestock health problems. Medical treatment should be given to sick flocks to keep their organic status. Growth hormones, genetic engineering, animal coning and antibiotics are not allowed as they are not organic. USDA-accredited inspection organizations can certify all organic poultry manufacturing and processing processes, such as egg processing facilities and slaughter plants. A meticulous record of all feeds, prescriptions administered, and payments should be maintained. To protect their purity, organic birds and poultry must not come into contact with prohibited chemicals or be combined with non-organic materials.

Organic production is defined as «a production system that is managed… to respond to site-specific conditions by integrating cultural, biological, and mechanical practices that foster cycling of resources, promote ecological balance, and conserve biodiversity."This ensures that in order to be qualified for organic livestock or poultry processing, farmers must follow ethnic, biological, and/or mechanical methods, as well as ecological laws such as natural resource conservation and resource use (Dippel *et al*., 2014). Some important factors to be considered before starting organic poultry production are;

- The soil in the chosen area should be relatively dry, as the wet ground can be difficult to reach and difficult for birds to navigate. It will also make it difficult for transport vehicles to gain access to the farm.
- A shelter is a must for poultry and exposed or open locations should be avoided since they pose a threat to the security of the poultry.
- Organic poultry farming requires intense labour than any other facilities and, therefore before starting one should have a readily available labour force.
- Facilities like water and feed should be provided in the brooder house at all times since during the rotations, houses may be far away from the farm and there could be a chance of going long distances in adverse weather conditions.
- One needs a good amount of capital investment. This is because money is needed for literally every activity on the farm from the purchasing of the chicks to maturity.

6.4.1 Organic System Plans

To be credentialed, all organic producers–including poultry producers–complete an Organic System Plan (OSP) that:

- Explains what they are producing and how they produce it.
- Lists and describes the materials they used during production and planned to be used for the operation.

- Explains the process of monitoring and assuring that the operation follows the certified requirements.
- Explain their record-keeping system;
- Explain measures taken to avoid contamination.
- Give all other information required by the certification agency.

OSPs should be modified every year, and certification agencies should be informed immediately of any alterations to the procedure that may issue its abidance with NOP demands, including the application of a banned substance to any field, facility, livestock, or product that is part of the organic poultry operation.

6.4.2 Transformation Period

The rules for transformation defined by the organic certifying administrative body are;

- There should be a simultaneous transformation of land and poultry, but the conversion of land has specific rules according to the organic society.
- The poultry that is being transformed into organic can be reared on the field that must have already gone through a transformation period of at least 12 months as stated by the organic society.
- The transformation duration starts with the day of the first inspection.
- If the field and poultry transformation does not occur at the same time, then the poultry must be reared for a specific duration of time as stated by the organic council before the products could be traded as organic. The minimum duration for raising meat poultry is starting from the second day of hatching till the entire lifetime and eggs for six weeks.

6.4.3 Feed and Feed Additives

All organic poultry must be fed 100 percent non-synthetic feed, according to Section 205.237 of the Last Rule. The term "feed" refers to "edible products that are used by livestock for their nutritional value." Concentrates (grains) or roughages (hay and fodder) may be fed in mash or pellet form. Pellets, as opposed to mash, are more useful and must be micro-sized. Fermented rice, dried fruits and vegetables, and other refined agriculture foods that are organic are all allowed as poultry feed.

The whole field should be organic in nature, and poultry goods would be accredited organic on their own if they were fed a limit of 85 percent for ruminants and 80 percent for non-ruminants with organic origin feed that met the board's requirements.

The word "feed" includes all agricultural resources including pastures eaten by livestock for nutritional purposes. To put it another way, all of the ration agricultural components must be certified organic. Certification is needed in all fields including pastures used for organic chickens and lots used for outdoor entry. Both farm-raised and imported feed and feed additives must be recorded. Crop growers who cultivate livestock feed for sale to organic poultry producers must be licensed under the $5000 organic exemption for small-scale farmers and cannot market their feed as organic (de Jonge *et al*., 2013).

It must be understood that supplements could be included in the feed only after getting approval from the board upon the advice of a competent veterinarian and should never contain any genetically modified substances.

The following are crucial details for using additives and processed feed;

- Substances used for packaging such as emulsifiers, stabilizers, surfactants should have an organic root.
- Natural anti-oxidants are allowed.
- Natural acids are allowed to be used as preservatives in feeds.
- Colorants, additives, odour masking chemicals and appetite stimulants can all be organic meaning they came from nature.
- Antibiotics, medicines, growth promoters and others should not be fed to poultry in an organic farm.
- Sea salt, yeasts, vitamins, starch, honey, lactic, acetic, formic and propionic bacteria are the only additives that can be used for crop residues.

Non-synthetic (natural) substances as feed additives and nutrients may be listed on the national registry, for instance, oyster shells, calcium carbonate, or fish feed and other synthetic substances. DL-methionine is the only product in the national registry of special synthetic feed for poultry. For organic poultry farming the domestic; the list contains the following maximum rate of synthesis of DL-methionine per ton of feed: laying and gridding chickens 2pounds, turkey and all other poultry 3 pounds. The list includes DL-methionine, DL-methionine analogue and DL-methionine hydroxyl analogue calcium.

Trace minerals and vitamins that have been approved by FDA are allowed in poultry rations such that they are not derived from slaughter by-products and do not contain any genetically modified chemicals referred to in the last NOP legislation as the "excluded processes". This is because in the national register no synthetic colourings, flavourings, dust suppressants or flowing agents are available.

Feed for processing organic poultry shall not contain;

- Growth promoting therapy like hormones.
- Feed vitamins or ingredients in doses that are greater than what is required for healthy nutrition and wellbeing.
- Rubber pellets;
- Urea or composite.
- By-products of mammalian poultry.
- Use of dietary ingredients or supplements not licensed by the food and drug administration.
- Antibiotics like ionophores have been added to feed or forage.

Figure 6.2: Poultry feeding in a poultry house.

Source:https://www.google.com/url?sa=i&url=https%3A%2F%2Fagronomag.com%2Forganic-chicken-feed-short-guide-on-the-nutrition-of-organic-chickens%

6.4.4 Living Conditions

Organic bird habitat standards differ in that they must comply with certain guidelines different from conventional methods. For example

- Caging of the poultry is not allowed.
- Poultry should have easy access to streams, ponds or other water sources available if the weather allows.
- The floor of the brooding pen must be a firm construction and must be covered with litter material like straw, wood shavings such which are dust-free and soft and sand and turf of the initial thickness of about 10cm. Organic standards also permit the usage of paper-based bedding.
- Manmade daylights are allowed in the poultry area for altering of day length but are limited to only a maximum of 16 hours per day (Donnik *et al*., 2016).
- The field should have open space for the poultry to run around and about freely.
- Plants on the field are of great importance as it poses as natural feeds to the birds and also water.
- If birds must be housed due to departmental certification restrictions, the appropriate roughage and other criteria should be permanently be included to satisfy the ethological needs of the birds.
- The smallest floor space should be at an average of 30 chicks per square meter for day-old birds and 10 chicks per square meter for 4 weeks old birds. It should be defined and maintained by organic standards and should have a maximum of three levels above the ground.
- If new flocks are added to the shelter or building, the building must be washed up and disinfected properly. The vegetation should be allowed to regenerate by leaving the runways free.

Organic poultry producers must have an attitude that promotes the bird wellbeing and natural behaviour all year round and preserve the environment. This should be accomplished by:

- All birds should have access to the outdoors during the year including shade fitness areas, fresh air, clean drinking water and clear sunlight suitable to the species, its stage of growth, the environment and the ecology.

- Healthy dry bedding must be of natural herbal roughage.
- Shelter built to accommodate:
  - natural upkeep, comfortable habits, and the ability to exercise;
  - temperature, drainage ventilation, and air circulation that are appropriate for the species
  - Ensure the risk of animal injury is reduced.

To prevent the flow of dung and untreated water into surrounding or neighbouring surface waters or overland borders, the yard, feed pads, feeding grounds and laneways for organic poultry should be monitored and kept in order.

Because of the following reasons, organic poultry farmers can provide temporary confinement or shelter:

- Adverse conditions
- The age of the animals.
- Situations in which the birds' fitness, protection or wellbeing can be compromised.
- Hazards to the consistency of the land or water.
- Preventive health care treatments or sickness or accident management.
- Animal sorting and transportation.
- It is essential to breed.
- Demonstration at trade shows or other activities.

Where it is determined that temporary confinement of birds is required to protect the health, safety or well-being of organic flocks, farmers and the certification board may cooperate to decide on an appropriate method and time of confinement without compromising organic certification.

The accrediting agent must allow the type of temporary containment to be used, and it must provide the required care for the birds while still satisfying the existing criteria of the NOP Final Rule. According to the NOP Program Handbook, continuous complete containment of any animal indoors is banned.

Please note that all organic farmers have to take steps under the ultimate NOP law to avoid runoff and reduce soil depletion. Organic poultry farmers must not cause flooding or contaminate their birds with water sources. Bio poultry operations must deal with manure in such a way that plant nutrients with heavy metals or pathogenic species avoid the loss of crops, soil or water and optimize the nutrition recovery process.

For new structures or replacements, organic poultry farmers do not use arsenate-treated timber or other banned products in contact with soil or birds. It does not refer to timber used for the use of fence stations or construction materials in close contact with organic poultry until the birds are separated by using electric clamps, netting or other measures authorized by the certification agent. If treated wood was present before registration, it may be maintained but no further installation is permitted in installations such as styles, runways or buildings where birds may swallow drillings directly around the posts or timing up against wood (Durmic & Blache, 2012). Untreated, rot-resistant trees may be used, including cedar, white oak or black lime and stains in metal or concrete. The certifier also would not challenge the use of paints, sealants, and whitewashes in barns or other living facilities.

The surface area of the coating should be smooth, well-insulated, and simple to purify and fan.

Figure 6.3: Chicken farmhouse.

https://www.poultry.care/wp-content/uploads/2020/09/production-1024x567.jpg.

6.4.5 Health Care Management

The key goal of poultry health is to deter diseases from occurring by adopting health protection principles. A licensed veterinarian tests the entire flock or batch of birds for any signs of disease at the start, and a health care plan is established. The values of health care include the selection of appropriate breeds that conform to the climate and atmosphere, as well as the establishment of adequate standards and an emphasis on sanitation in the farm sector.

- To establish a natural immunological defence, the poultry birds should be fed high-quality feed and have access to routine open-air runs.
- The primary goal of curing an illness is to relieve the animal's pain and discomfort, even though this necessitates the use of drugs that do not meet ethical requirements.
- The vaccinations used on organic chickens should be approved by the organic council first.
- Unless recommended by a licensed veterinarian and provided under his or her care, underperforming poultry or birds with diseases will most likely be treated with homoeopathic or ayurvedic items rather than antibiotics.
- It is not allowed to use allopathic medicines or antibiotics to prevent diseases by increasing production.
- Under the supervision of a medical expert, hormonal procedures for cancer control can be carried out. Growth stimulants, on the other hand, are specifically banned in organic poultry. Such participants, on the other hand, use so to boost their growth and fattening, which is incorrect in several respects.

Organic poultry farmers shall establish preventive health care protocols under Section 205.238 of the Final Rule.

- Choosing species and varieties of poultry that are appropriate for local environments and immune to diseases and parasites.
- Provision of a nutrient-dense feed ration that includes vitamins, nutrients, protein and/or amino acids, fatty acids, and energy sources to fulfil dietary requirements.
- Establishing adequate housing, grazing conditions, and sanitation activities to reduce disease and parasite incidence and spread.
- Physical improvements are needed to improve the health of poultry performed without discomfort and stress.
- Vaccines and other veterinary biologics administration.
- Conditions to allow for proper training, the right to travel and the management of stress (for instance, no laying caged hens).

6.4.6 Cleaning and Disinfection Products

Cleaning and disinfection materials should be certified by the Organic Certification Body. Potassium and sodium soap, lime juice, lime, quick lime, alcohol, formaldehyde, natural plant essences, caustic potash, sodium carbonate, water, steam and hydrogen peroxide are among them.

6.4.7 Physical Modifications

As previously mentioned, physical modifications are permissible provided they are carried out in a way that promotes the health of the poultry while minimizing discomfort and stress. Before making physical changes, organic growers can consult with their certifying agent to see if the process is permitted and provide a description of the procedure and the reasons for its inclusion in their Organic System Plan.

Physical modifications such as the ones mentioned below are usually permitted if the conditions outlined above are met:

- Wing clipping or toe clipping for turkeys.
- Hens for laying eggs: Trimming the beak or the wings.

Changes proposed: On December 2, 2011, the National Organic Standards Board (NOSB), which advises the USDA on the NOP Regulations' enforcement, issued a resolution barring the following: I De-beaking, de-snooding, caponing, dubbing, and toe trimming turkeys; (ii) Toe trimming turkeys unless infrared in the hatchery; and (iii) Beak trimming turkeys unless they are under 10 days old. The USDA has disobeyed the NOSB's advice.

The NOSB also recommends that: (a) ammonia levels are less than 10 ppm outside and less than 25 ppm indoors; (b) cage containment of birds be prohibited at all times, and (c) minimal indoor and outdoor room standards for organic birds be defined.

The NOP should define the following avian low space specifications, according to the NOSB guideline dated 12/2/2011:

- Laying hens and breeders need 2.0 square feet per bird indoors and 2.0–5.0 square feet per bird outdoors.
- 2–3 lbs/sq ft within, ft outside. pullets 2–3 lbs/sq ft indoors
- 1–5 lbs/sq ft indoors, 2–5 lbs/sq ft outdoors for broilers
- Breeding, laying, or meat turkeys and geese (pounds): The final guideline makes no mention of this.

- Meat from ducks:
- Ducks are eggs laying.
- Breeder ducks.

6.4.8 Medications that have been approved by the organic board.

In the case of health insurance, the Final Rule states in 205.238 that, "when preventive practices and veterinary biologics are inadequate to prevent sickness, a producer may administer synthetic medications, provided that such medications are allowed under §205.603 (the National List)." The use of such prescriptions can be subject to stringent cessation or "veterinarian prescription-only" requirements, and failure to adhere to these requirements may result in credential loss.

According to Section 205.238, an organic poultry production does not market, mark, or reflect as organic any animal or edible commodity originating from an antibiotic-treated animal, any material containing an artificial substance not authorized under 205.603, or any substance containing a non-synthetic product banned in 205.604.

Animal medicines, other than vaccinations and other biosimilars, are also banned in the absence of disease; growth-promoting hormones; synthetic parasiticides daily; parasiticides for slaughter; and the use of animal medicines over the Food, Narcotics and Cosmetic Act.

Organic chicken farmers are not allowed to "withhold veterinary attention from a sick animal to protect its organic status," according to Section 205.238(c) (7). When organic farming processes malfunction, the available treatments must be used to regain an animal's health. Livestock that has been processed with a banned drug must be specifically marked and cannot be marketed, labelled, or portrayed as organic."

National List

On the National List, disinfectants such as ethanol and isopropanol are allowed, as well as aspirin, atropine, butorphanol, flunixin, furosemide, magnesium hydroxide, vaccines and biologics, poloxalene, tolazoline, xylazine, chlorhexidine for surgical procedures and as a teat dip; chlorine materials and peroxyacetic/peracetic acid as sanitizers (Escribano, 2016).

As previously mentioned many of the drugs described above have limitations on their use or require lengthy storage periods. Before using a drug, make sure it complies with any restrictions or annotations mentioned on the National List. Be sure the drug is on the Organic System Plan, and keep track of all applications. Check with your certification body if you have any questions about the condition of a certain brand or manufactured substance before purchasing or using it.

A short list of natural compounds that are banned in organic livestock processing can be found in Section 205.604 of the Final Rule. Strychnine is the only thing on the list right now.

Excipients

Excipients are included in drugs given to organic poultry, but only if the excipient is: FDA-identified as Generally Recognized As Safe (GRAS); FDA-approved as a food additive; or included in the FDA evaluation and acceptance of a New Animal Drug Application or New Drug Application.

Excipients are: "any ingredients that are intentionally added to livestock medications but do not exert therapeutic or diagnostic effects at the intended dosage, although they may act to improve product delivery (e.g., enhancing absorption or controlling the release of the drug substance). Examples of such ingredients include fillers, extenders, diluents, wetting agents, solvents, emulsifiers, preservatives, flavours, absorption enhancers, sustained-release matrices, and colouring agents."

Parasiticides

Fenbendazole (CAS # 43210-67-9), ivermectin (CAS # 70288-86-7), and moxidectin (CAS # 113507-06-5) are the only parasiticides currently on the National List. Parasiticides can only be used on: (1) Breeder stock before the third trimester of pregnancy but not during lactation for progeny that will be marketed, branded, or represented as organically produced; and (2) Dairy stock before the processing of milk or milk products that will be sold, labelled or represented as organic. In the manufacture of organic eggs and foods, parasiticides are not permitted.

Birds slaughter

According to organic guidelines, there should be minimal discomfort and pain during the killing of poultry.

The poultry must be washed and decontaminated with organic council-approved materials.

To meet all of the organic council's standards, poultry goods must be packaged in a sanitary manner. Chemicals are not permitted to be used during the packing process.

Washing and disinfection should have separate rooms

There is a legal age for slaughtering various species of chickens, and the poultry cannot be killed before they reach that age. Rapidly growing strains, for example, can be slain after 81 days, while sluggish growing strains can be slain after 70 days. Slow-growing chickens achieve less than 45 grams a day in live weight.

6.4.9 Organic Poultry Product Processing and Labeling

Poultry slaughter and egg storage and processing processes must be accredited to be branded organic. During handling or refining, organic goods must not be mixed with non-organic items or actually interact with banned chemicals. Both additives and other materials used in or on organic poultry products during production must be listed on the National List in section 205.605 or 205.606. Data with all manufacturing operations must be kept.

The brand marking should be legitimate and have a product code or date code that can be traced back to the poultry or project from which it came. The USDA Organic seal can be used by all operations approved by NOP-accredited certification organizations on goods that have at least 95 percent organic content and follow all relevant labelling criteria, though it is not required (Escribano *et al*., 2019).

6.4.10 Transportation of Organic Poultry Products

Great care is supposed to be taken when transporting poultry products. One should ensure the birds are not in any stressful condition, hungry, thirsty, malnourished, injured and in pain which may cause them discomfort.

Several rules are governing the transportation of poultry:

- Proper arrangements should be made before the time of travel to satisfy the requirements of the birds.
- One must ensure the birds are in good health during transporting.
- The delivery process can be done in such a way that does not harm the poultry.
- During the flight, skilled employees should be assigned to manage the poultry.

- The poultry must be monitored on regular basis at short intervals.
- There should be enough space for the birds to travel in comfort.
- During the shipping path, water, food, and rest must be given at periodic times.
- Tranquillizers and electrical stimulators are not allowed to be used when boarding or off loading birds.

6.4.11 Breeding Management

In the poultry industry, organic farming practices do not only depend on food and feed, but also on respecting and protecting the animals.

The certification board's reproduction guidelines for organic chicken farms are as follows:

- The breed range should be suitable for the local circumstances.
- Natural reproduction is preferred, but artificial insemination may be used in exceptional circumstances.
- Genetic engineering-based breeding methods can never be used on organic farms.
- Hormonal therapies can only be used if they are medicinal and are intended to address physiological issues.
- Mutilation procedures such as beak thinning are not allowed until they have been approved by an authorized agency for protection or to improve the welfare of the birds. Surgical procedures should only be performed at the required age to minimize pain and discomfort, especially in older birds.

6.4.12 Waste Management of Organic Poultry

There is a lot of waste that comes from organic poultry and should therefore be well managed and properly disposed off. The waste should be managed such that;

- Soil and water erosion are kept in check.
- No nitrates, phosphates, or microbial pathogens are released into the environment.
- Manure can never be contaminated on the farm, and the density or rate of applications are determined by approved accreditation bodies depending on the farm's local circumstances.
- The most nutritious recycled materials are collected.

Careful storage, recycling, and treatment of poultry manure are needed to avoid pollution of the land and surface water.

On organic fields, practices such as incarceration are banned because they can lead to pollution of both the soil and the air.

6.5 COMMON DISEASES AFFECTING ORGANIC POULTRY AND THEIR TREATMENT

Marek's Disease

This disease is a viral disease that is highly contagious in birds.

Symptoms

- Paralysis in the legs, wing, and neck
- Lameness
- Weight loss
- A grey pupil in the eye
- Blindness

Organic medications for 50 birds

- Aloe Vera gel 45g
- Black pepper 45g
- Cayenne pepper 48 g
- Water 0.5 lit.

These should be given to the birds for 5days

Infectious Bursal Disease (IBD)

This disease is also known as Gumboro it's also highly infectious.

Symptoms

- Loss of appetite
- Whitish, watery poop
- Weight loss
- Paleness, droopiness and huddling

Treatment for 50 birds

- Tagiri extract. 30 ml
- Aloe Vera gel 50 ml

- Black pepper 45g
- Cayenne pepper 45g
- Water 0.5 lit.

Newcastle Disease

Symptoms

- Coughing and sneezing
- Respiratory distress and nervous disorder
- Laying down in young birds
- Abnormal positioning of the head
- Death occurs within 3-4days

Treatment for 50 birds

- Tagiri extract. 30 ml
- Cayenne pepper. 45g
- Wood Ash. 8 g
- Aloe vera gel 45g
- Water. 0.5 lit.

E. coli

Symptoms

- Ruffled feathers
- Yellowish coloured droppings
- Soiled vent
- Decreased appetite

Treatment for 50 birds

- Ginger 25 ml
- Garlic 25 ml
- Scent leaf 25 ml
- Water 0.5 lit.
- Give these for 5days

Diarrhoea

Symptoms

- More water in the bird's droppings

Treatment

- Bitter leaf
- Scent leaf

How to use

Blend these leaves with 0.5 lit. of water and pour into their drinker (give these for five days)

Coccidiosis

Symptoms

- Ruffled feathers
- Loss of appetite
- Droopiness
- Pale Combs
- Blood in their poop

Treatment

Blend some amount of bitter leaf with 0.5 lit. of water and give the birds to drink for five days

Fowl pox

Symptoms

- Dry pox: decrease in weight gain
- Dark brown/yellowish like lesions on the head
- Wet pox: lack of appetite
- Nasal and ocular discharge

Treatment

- Administer organic multivitamins and molasses added with bitter leaf extract for 5days

Note: Endeavor to feed the birds well and adequately during this period of fowl pox

Chronic Respiratory Disease (CRD)

Symptoms

- Coughing and sneezing
- Decreased feed intake
- Low body weight

Treatment for 50 birds

- Ginger 20 ml
- Cayenne pepper (dry) 20 ml
- Water. 0.5 lit.

Worms

Symptoms

- Worms in the bird's poop
- Low feed conversion rate

Treatment for 50 birds

- Ginger 20 ml
- Garlic 20 ml
- Pawpaw leaf 30 g
- Water 0.5 lit.

Grind the pawpaw leaf with the ginger and garlic in 0.5 lit. of water give to the birds for five days

Infectious Coryza

Symptoms

- Sneezing
- Swelling of the face

Treatment for 50 birds

- Ginger 20 ml
- Garlic. 20 ml
- Tumeric 20 ml
- Tagiri extract 30 ml
- Water 0.5 lit.
- Administer for 5days

Heat Stress

Symptoms

- Dropping or spreading of wings
- Panting (opening and closing of the mouth)

Treatment for 50 birds

- Drop ice cubes into their water
- Feed them during the cooler hours of the day, do not feed them during the hot hours of the day
- Avoid overcrowding.

6.6 ADVANTAGES OF POULTRY FARMING WITH ORGANIC PRODUCTS

Organic poultry farms are usually cage-free, allowing the poultry to roam freely. This reduces the labour and feeds requirements.

Because of the agricultural techniques used, organic poultry farming increases the ecosystem of the region where the field is located.

There is no or less use of chemical additives, therefore, makes the products be most preferred by people because they are considered healthy and friendly to the body.

The poultry' taste or product consistency is superior to that of other customarily reared chickens. This is because they are fed only organic feeds with no chemical additives.

Since cultivation and poultry are interdependent on the same field, there is a low expenditure.

6.7 LIMITATIONS OF ORGANIC POULTRY

Many people lack knowledge about organic poultry farming and this leads to the need of having trained expertise around. This further increases the cost of production as related to the profit as some farmers are scammed by middlemen.

Poor infrastructure. Many countries have poorly developed infrastructure that inconveniences the activities under organic production. For example, financial aid and certifying agencies, poor roads and lack of electricity (Entz *et al*., 2009).

Very many stringent hygiene and quality regulations enforced by developed countries make it difficult for developing nations to sell their goods. The decrease in the price of the products in the world market also makes farmers lose their morale.

There are fewer training facilities for poultry farmers.

Remedies

Create awareness among people about the need to eat healthy and organic food. This provides a market for organic poultry products both at the national level and internationally.

Improve infrastructure. The government should ensure there is enough financial aid to the farmers. This can be in form of loans at low interest rates so that they are affordable. The conditions of roads should also be improved to facilitate easier transportation of products without delay which may cause perishing of some of the products. Electricity should also be distributed to all households even in rural areas.

Provide training in organic poultry farming. The government should establish many vocational training centres for farmers interested in gaining knowledge about organic poultry and those students who fail to merit to colleges should be encouraged to join such institutions.

The government should work at creating a market at the national level instead of always depending o exports to developed countries. It should also create good relationships with other countries to enhance trading activities.

Chapter 7

Organic Beef

CONTENTS

7.1 INTRODUCTION

One common livestock product is beef. Organic beef has been obtained from an animal that is fed on organic feeds. Therefore, organic principles have been used in rearing cattle. However, there are several factors that should be taken into consideration for livestock to be termed as organic. There are standards put in place by the National Organic Program. These standards outline criteria to be met for an animal to be certified as organic cattle. They also make use of information on the history of the animal such as its breeding history, its feeds and medication utilized during veterinary visits. This information is needed before cattle are termed as organic. In most cases, organic beef is said to be obtained from animals that have not been given growth hormones or any form of antibiotics. Animals' from which organic beef is obtained should not be fed on food having chemicals. No pesticides should be used in controlling pests. Certifiers usually inspect the farms from which these animals are reared to ensure that farmers have met the National Organic Program standards. Companies managing the production of organic food need to be evaluated and certified. This is because methods used in the production of organic beef are different from the production of other beef (Escribano, 2016).

Figure 7.1: There has been a growing demand for organic beef products.

Source: https://www.kingsoopers.com/p/simple-truth-organic-90-lean-grass-fed-ground-beef/0001111096784

7.2 CRITERIA FOR CERTIFICATION OF ORGANIC CATTLE

Genetic engineering should not be used in producing these animals. Organic cattle should carry original genes from the previous cattle. Therefore, the genetic composition of the animal should not be altered. Also, the animal should not have come in contact with any form of radiation more so ionizing radiation. The animal should not have consumed any form of sewage sludge. The animal should continuously access the outdoors. This is for animals raised in systems such as zero-grazing. They should be allowed to get sun while rooming outside. Access can be denied in cases where the weather is poor and could negatively affect the health of the animal. During veterinary care, the animal should not be given antibiotics or growth hormones.

Regarding feeding, organic animals should be given supplements meeting required standards. This implies that all the minerals or components prohibited (Manure and urea) should not be given to these animals. They should be given organic feeds and in the case where they are unavailable then they should be given recommended organic by-products. They should be reared on land meeting the organic crop production standards. During gestation, from the third trimester, the animals should be managed organically. These standards seek to ensure that these animals are as natural as possible (Gunnarsson & Mie, 2018).

There is a distinct variation between organic beef and conventionally produced beef. The variation is due to the rearing and handling of these animals. Most boards overseeing the production of organic been have not made any claims stating that organic foods are more nutritious or safe compared to other foods. However, they have developed labelling rules that enable customers to know the content of the food. Most beef products having the USDA Organic Seal are known to have labels. The seals are utilized by certified farmers.

Figure 7.2: The USDA Organic Seal is used in proving a farmer is certified

Source: https://www.clemson.edu/public/regulatory/organic/seals.html

7.3 PRINCIPLES OF ORGANIC BEEF PRODUCTION

These principles are outlined in the Principles of organic livestock and livestock products. These principles need to be adhered to in the production of organic beef. Some of the general principles are as follows:

In areas where livestock production is done, there is a need for an organic farm unit. They should also be maintained according to the organic farming guidelines. Organic livestock farming should be done in a manner that positively impacts the organic farming system in that it should help improve soil fertility, improve biodiversity and help properly manage flora by utilizing good grazing systems. For animals raised entirely on grass, farmers are required to ensure they have good access to pasture or the required forage. They also need to ensure that animals enclosed in systems can access the outdoors. This will help improve the animal's physiological health. In cases of poor weather, the farmer needs to ensure that proper structure has been put in place. This will protect these animals from harsh weather preventing the incidence of some diseases in cattle. This caters to the welfare of the animal.

Organic farming principles require the farmer to have knowledge of livestock stocking rates of their region. This is information is vital at the starting phase as they can approximate the capacity of feed production, nutrient balance, animal health and the possible effects on the environment. They need to ensure the management practices used focus on utilizing organic methods of farming. This implies that all activities involved in livestock rearing such as breeding, pest and disease control, feeding and health care provisions should be natural. In the case of feeds, they should be organic and meet the required standards (Grunert *et al.*, 2018).

When selecting animal breeds and breeding methods, they all need to be compatible with the principles of organic farming. This means that any form of genetic engineering should not be used. When selecting the breeds, the farmer needs to take into consideration their ability to adapt to the conditions of the area, the ability of the animal to be resistant to diseases and the likelihood of a particular animal breed suffering from a certain disease, for animals siring children the principles require them to be raised under this system for their entire life. Therefore, beef cannot be termed as certified if the animals in question have transferred from a different farming system to an organic farming system. However, there are certain cases where animals that have not met the requirement can be used in organic farming. These are exceptions put in place by the board to allow continuity. The animals can be used when there are expansions to be done on the farm. It is also done in the case where there is new specialization regimes put in place to allow the development of organic farming. It is also used when the herd needs to be renewed. This is when there was a catastrophe that led to the death of animals. The exceptions are also applicable in the case where the males are to be used in breeding. For farmers who have converted their farms from nonorganic to organic, some conditions should be met before their beef products are certified as organic. They should convert their pastures and feeds to be more organic and comply with the regulations. There is a certain time given to farmers to conduct the conversion. These durations may vary depending on the farm. The Organic Farming Programs may also change the durations outlined for the transition of the farms.

With regards to nutrition, farmers need to ensure that the systems they are using feedstuffs giving animal's optimum amount of nutrients. The feedstuffs however need to be in accordance with the regulations. The implementation of feeding guidelines the relevant authority needs to ensure that livestock products produced by farmers can maintain the needed organic standards. They also need to meet the organic standards put in place by the relevant authority. The feeding rations given to certain animals need to be accounted for. They need to ensure feeds are given to animals according to their ages; when they are young or nurturing (Honkanen *et al*., 2009). This is because, at this stage, they need more energy and more amounts of food. Before the board gives the guidelines they need to be permitted by the national legislation. They regulate the feeding of animals. Farmers need to give foodstuffs that will maintain good animal health and welfare.

The guidelines on feedstuffs require that plants given to animals have not be sprayed with any form of chemicals or chemical solvents. They should be of mineral origin and contain minerals. These can be utilized if they are of natural origin. In the case of a shortage of feedstuffs, farmers are allowed to use organic by-products. Farmers are allowed to use additives obtained from natural sources. This applies to anti-caking agents, surfactants, binders, thickeners and emulsifiers. They should also use preservatives made using natural acids. They should be colouring agents, appetite stimulants, antioxidants obtained from natural sources. For farmers who feed their animals on silage, the additives and preservatives used should not be obtained from genetically engineered products or by-products. Therefore, they may have yeasts, by-products of sugar such as molasses, sea salt, enzymes, honey, natural acids and natural bacteria.

With regards to health care, organic livestock principles require organic methods to be used in the prevention and treatment of diseases. For any animal experiencing growth challenges, growth hormones should not be used to boost animal growth. Farmers are encouraged to properly identify the breed of an animal less likely to suffer from animal diseases. They are advised to keep animal breeds having strong resistance to animal diseases. Regulations require farmers to allow animals to get exercise in open runs and eat good quality food. This will enable animals to develop natural immunity against diseases. They need to ensure good feeding systems have been utilized; adequate numbers of cattle should be kept in farms to minimize the spread of livestock diseases.

7.4 RAISING ORGANIC BEEF CATTLE

Organic beef is obtained from organically reared cattle. There are different types of reared cattle reared organically. There is grain-fed beef and grass-fed beef. For organically reared beef, they are fed on grass. They are therefore grass-fed cattle. Raising animals of grass is more beneficial for farmers as they do not have to spend purchasing feeds for their animals. They can grow it, harvest and dry it for later use as hay. In the case where the farmer is unable to grow their own grass, they can purchase hay. This is less expensive compared to grains. Grass feeding cattle is more beneficial to farmers compared to other feeding methods. Raising beef animals is a commercial activity therefore the farmer needs to adopt strategies that will enable them to run the farm smoothly. Over the years there has been a growing demand for grass growing cattle. There are many things that a farmer needs to consider before they start grass-feeding animals. They need to consider the type of grass they will feed their animal. This is because the quality of grass given to the animal will influence the quality of meat. They need to grow grass that has high nutrient content and has high Brix. The grass also needs to be fast-growing for the farmer to ensure the availability of food for the farmer (Ivemeyer *et al.*, 2012).

7.4.1 Forage Quality

For grass-fed livestock keeping, the quality of the forge is important as it will influence the animal's energy supply. This in turn influences the animal's weight. When the animal is given enough energy then they are likely to gain more weight approximately 2.5 lbs. in a day. The farmer needs to ensure that the forage given to the animal has a digestive content higher than 65%, crude protein of between 14-18%, water-soluble carbohydrates higher than 20% and dry matter higher than 20%. They need to have good nutritive content to ensure that the animal gets all the needed nutrients. The farmer needs to ensure that the animals gain about 2-2.5 lbs. daily so that the animal is on the right track. They can do so by ensuring that the forage given to the animal has monitored nutritive value. The forage quality should be properly managed.

Forage utilized in livestock production has different nutrient values determined by how the farmer manages the forage. The stage of harvesting will also influence the nutritive level of food. This is evident in hay and silages. Hay is grass-grown and dried before it are harvested while silage is grass harvested when they are still succulent. These two types of forges have different nutritive values. The farmer needs to manage them properly to ensure the nutrient content is well maintained. When managing forages, an important factor to note is that forages that have not reached maturity are more nutritious than those that have attained maturity. Once a forage plant reaches the reproductive stage the forage quality reduces.

Apart from forage quality, the farmer needs to take into consideration the amount of forage consumed by the animal is also important. Animals need to be allowed to eat the maximum amount of forage they can consume. At no point should the farmer stop the animal from grazing. The farmer, therefore, needs to have good management strategies to allow them properly control the forage intake. This will prevent any future pasture shortage which may result in animals lacking food. Most farmers opt to plant cool-season perennial forages. This is because they provide more permanent food solutions compared to other forages. Among the strategies used in forage management include animals being allowed to graze until a certain level. In most cases, they are not allowed to graze to a height less than four inches. In the case of warm-season forages, the grazing height is different from cold-season crops. In their case, they should be fed to a height of eight inches. The height levels should be taken into consideration when harvesting grass for hay or silage. This is done to allow the plant to have sufficient leaf mass to facilitate photosynthesis. This ensures the plants will re-grow.

Some of the species of cold-season forages include timothy, tall fescue, orchard grass, smooth brome, perennial ryegrass, red clover, reed canary grass, Alfalfa, annual grass, cereal rye, barley and wheat. Warm-season forage species include big bluestem, Indian grass, switchgrass and Gama grass. To better manage pastures, farmers are advised to carry out rotational grazing. This involves animals feeding in a particular portion of the field at a time. They graze on the forage till the required level then they are moved to new ungrazed areas. This feeding method is beneficial as it will reduce cases of selective grazing. This is important because selective grazing is known to negatively affect forage production. This is because some forages will not be fed on causing them to lose their nutritive value. Selective grazing is also known to cause forage proliferation (Jespersen *et al.*, 2017).

As grazing seasons cannot be ensured throughout the year, farmers need to ensure forage can be supplied. Forages should be properly harvested at the right time frames. Most farmers have harvesting schedules that should be adhered to. In cases where the land situation is not suitable for planting perennial forages, farmers can plant annual crops. Recent developments involve the growing of no-till crops such as cover crops to ensure there is a continuous supply of forage. There has been a growing dependence on an agronomic standpoint. This has proven to be beneficial for most farmers. In as much as raising grass-fed beef is a better route to take for farmers, they need to ensure they have all the needed resources before they can make any decisions.

Grain-fed organic beef

There are organic beef cattle raised on organic grains. Many grains could be fed to cattle. However, in the case of organic beef, the grains have to be organic. Among the major grains given to animals is corn hence the name corn-fed organic beef. This type of feeding is mostly found in the U.S. This kind of feeding is known to result in the production of very juicy and tender beef products. This type of feeding requires the animals to be fed in the pasture as they are given grain. This is one way of supplementing their diet in that they can get some needed nutrients by feeding on grains. In most cases, farmers allow the animal to feed on pastures till a certain age before they are moved to feedlots. In the feedlots, their diet is mostly composed of grains.

While in the feedlot, the farmer needs to ensure they are given a balanced diet wholly organic. Therefore, they may be fed on soybeans, corn, by-products, required vitamins and nutrients and forage. The feeds are to be given in appropriate rations to enable proper feeding. When fed properly, grain-fed organic beef produces good quality beef. The grains need to meet the animal's nutrient requirement. The animals also need to be given good care and adequate water supply (Kleter *et al.*, 2018). Farmers opt to use grains to feed their animals because there is certain duration where they are unable to ensure the availability of grass or forage.

There have been concerns that grain-fed beef may not be organic. This may not be entirely true as there are animals fed on purely organic grains their entire life. The supplements given are also entirely organic. Like grass, farmers need to take into consideration the types of grain they give to their animals. Grains such as corn are said to be highly nutritious. They are also rich in energy. This is beneficial for the animal as energy is needed to enable the animal to meet the 2-2.5 lbs. daily gain. It helps farmers ensure consistency and in addition, it results in the production of good-tasting beef. Another advantage of feeding grains to cattle is that grains are easily digested. This implies that animals can easily attain the slaughter weight and at a shorter duration of time.

The challenge of using grain to feed beef is that for grains such as corn, when they are fed to animals in high concentrations, they have certain health implications on the animal. In high amounts, corn can lead to the animal suffering from cow acidosis. This is a condition where the cow suffers from heartburn which may eventually result in the death of the animal. They also produce a high level of fat which when consumed by human beings will result in certain health complications. This is because excessive fats accumulate around vital organs and blood vessels restricting blood flow to body organs.

7.5 CERTIFICATION FOR ORGANIC BEEF PRODUCTION

In different countries, there are different procedures employed for a farmer to be certified. First, the farmer needs to comply with the rules and guidelines outlined by the Certification board. When the farmer has met the requirement they fill in an application form. The application for the farmer is required to give more information about the farm. Some of the information required is as follows:

The farmer should mention the type of farming system used in rearing livestock and the type of production they are using. They will be required to fill in their personal information providing detail such as their name, the contact persons, addresses and telephone number among others. This is vital information that would be included in the certificate. The farmer is also required to give their professional detail. For farmers who had initially applied for certification, they need to provide the name of the Certification Body they had applied (Kongsted & Sørensen, 2017).

As the farmer applies for certification, they will be required to pay a registration fee, travel cost and inspection fee. This should be paid as the application form is being submitted. Information on these fees is usually notified given to the farmers as they are given the application form. If any other information is required, the farmer will be notified. The farmer needs to ensure that they are applying for certification with the right body. Once the application form has been completely filled, the operator submits the form to the body of their choice. The body will then provide a standards document and an operation document. This outlines the requirement of the body to be met as the farmer rears organic livestock.

Farmers and the Certification Bodies agree on the roles to be played by both parties. This is followed by the farmer or operator paying the required fees to the relevant certification body. After all, this is done; the application document will be audited and given to the farmer. From here, the body will accept or reject the application. When the application is rejected, the form will be returned to the farmer listing the reasons why the application has been rejected. Some certification bodies allow farmers to reapply when they have met the required conditions. If the application has been accepted, inspectors will be scheduled to conduct farm visits. The farmers are given the scheduled dates to allow them to prepare for the inspection. The farm should have met the standards given by the board. The inspection is done by an external inspector. They are required to conduct an elaborate inspection of farm facilities and lands. They also determine whether the operations done on the farm comply with the board's standards.

The inspector will also verify the farmer's documents. Once the inspection is done, the inspector is required to give feedback to the certification board. This is done by giving an inspection report. When the report indicates compliance with Organic Farming Standards, the farmer is awarded a certificate. A summary of the inspection is as follows: inspection time is given to a farmer to gauge their compliance with the Board's standards. Compliance with the standards is important as it will determine whether the inspection will be done or not. In the case of compliance, an external inspector is sent to the farm to inspect the fields. An elaborate inspection of the land and farming operations is done. The farmer is required to allow the inspector to have access to farm records. The inspector should be allowed to access farm accounts from which he or she will calculate the farm's input and output norm. The inspector is required to develop an estimation of farm production. From that, the inspector should evaluate the production system. He or she will be required to make inquiries on the farm. Later on, he or she will be required to conduct a risk assessment. For farmers using off-farm inputs, the input will have to be assessed (Krieger *et al.*, 2017). The inspector will have to check whether the farmers have used any chemicals. To achieve this, the inspector will have to set samples of animal residue. These samples are sent to the lab for analysis and the results are included in the report.

There are a number of benefits farmers enjoy once they have been certified. For example, once the farmer has a certificate farmer may be eligible for loans. They are also given adequate time to pay their loans. Efficient payment plans are made for farmers this way they can pay their loans in a manner that will not affect farming operations. Another benefit that these farmers enjoy is the ease of exporting their beef products. Taxes imposed on these products could be reduced to allow easy transportation. Certified farmers also enjoy marketing benefits in that the body they are registered under can market their products. These farmers enjoy subsidies given to farmers under government schemes. Having a certificate makes the farmer eligible for any government programs or schemes. Certification of farmers could be accompanied by farmers registering under other boards dealing with food production. For instance, in India, once the farmer has been certified they can proceed to be registered under the Food Safety and Standards Authority. This body is known to handle and monitor food businesses in different countries. The board is involved in the manufacture, processing, storage, marketing, selling and distribution of farm products. In some countries, the farmer may be fined if they are not registered under these boards. Farmers must be registered under these boards as they have information on the market and efficient strategies to be used in marketing these products. Registration involves the farmer filling an application form and paying certain fees. Once accepted the farmer will be notified. Generally, Organic farmers need to meet the standard laid down by the APEDA. APEDA is the body that regulates other certification bodies. It is the body in charge of creating rules and regulations followed by organic farmers.

7.6 HEALTH OF ORGANIC LIVESTOCK

In organic beef production, one crucial element is the health of the animal. In organic beef production, animal health and welfare is highly emphasized.

This is why there are some regulations put in place to ensure animals used in organic beef production are healthy. Animal health is also needed for a farmer to be certified. The EU council Regulations have set rules to ensure that animal immunity is effective, it improves animal welfare, minimizes damages to fauna and reduces meat residues. They are concerned with consumers getting good quality products. Under animal health is animal welfare.

To ensure animal welfare, Some principles specifically deal with animal welfare. They are listed under human control. Animals should be given freedom from thirst and hunger. Farmers need to ensure animals can access good water and food. The food should meet all animal dietary requirements and should be a balanced diet. This will ensure good animal health. The rules also state that the cattle should be given freedom from discomfort. This implies that animals should be raised in suitable conditions. They should have resting areas and be sheltered from harsh weather. Animals should have freedom from disease, pain and injury. Farmers need to ensure that animals are protected from diseases. Proper diagnosis should be done on the animal and organic methods of treatment used.

Animals should be given the freedom to behave normally. This is achieved by providing animals with the needed space they require. The proper company should be provided for animals to enable those exhibits natural behaviours. Animals should be given freedom from distress and fear. Farmers should ensure animals are raised in an environment that will not cause poor mental health. To ensure good animal health, farmers should take necessary measures to ensure animals are protected from diseases (Leeb *et al.*, 2014). For organic farms, the prevention of diseases can be achieved through a number of methods. The farmer can choose to use animal breeds having high resistance to diseases. They should identify animal strains resistant to diseases. Evaluating the level of biosecurity can help prevent the incidence of diseases in animals.

Proper routine practices should be done on animals. This applies to animal husbandry. The practices should be specific to a certain animal. Animals should be given high-quality feeds and should be allowed to feed at any time enabling them to develop natural immunity against diseases. The number of animals reared by farmers contributes to the incidence of diseases. Farmers are advised to keep appropriate numbers of animals. Overgrazing will lead to a higher probability of animals suffering from diseases. Overgrazing makes the spread of diseases faster and significant damages could be incurred. Animal health does not entirely imply that animals should not suffer from diseases. It covers treating ill animals to prevent the spread of diseases to other animals.

Medicines used in organic farming are not the same as those used in conventional farming. There are certain drugs prohibited in organic beef production. Organic beef production systems should be made in a manner that allows proper pest and disease control. These systems aim at ensuring there is minimal dependence on medicines and improve the health and welfare of animals. Among the plans, farmers should have is the animal health plan. A health plan requires the farmer to have an organic livestock record book. This book entails all practices done for animals to ensure good health. For farmers converting from other methods of farming to organic farming, they need a health plan developed by a veterinary officer.

The health plan covers all the health problems experienced on the farm. A record book helps to narrow down the specific problems experienced on the farm. It also defines methods to be used by farmers in solving health problems.

Health plans should tackle the following issues: identification of diseases already present on the farm. Effective solutions are useful in solving identified problems. They also need to develop modifications to be implemented in the farm to help the incidence of diseases in the farm. This includes the adoption of rotational grazing to control the spread of diseases. To identify animal diseases, their faeces are tested. It can also be used to establish the presence of internal parasites. Farmers need to ensure their health plan is up-to-date. They need to look at the future implications of using the treatment. There are several steps the farmer can take to come up with a health plan.

The farmer should access the vet for him or her to draw the health plan. The farmer then proceeds to identify the disease. They get information on the life cycle of the identified parasite or health problem. Once that is done they can look for available treatment options adhering to the regulations. They also need to be aware of current treatment options. They then develop strategies useful in breaking the life cycle of the pest or disease. They gradually reduce the animal's dependence on medicines. Farmers then proceed to adopt good management practices. If the management practices are not successful in controlling animal diseases, the farmer should look for advanced medicines. In as much as they should be effective, the farmer needs to ensure that they are organic treatment methods. A good health plan should indicate the withdrawal period for the used treatment (Lu *et al.*, 2010).

Prevention of animal diseases requires routine management practices such as dehorning to be carried out. These activities should be done in compliance with the Organic Certification Board guidelines. They require little or no chemicals to be used during these practices. There are special cases where the farmer is allowed to use synthetic medicines. In situations where the animal is experiencing much pain, synthetic medicines can be used to relieve the pain. The farmer needs to detect the disease while in its initial stages to enable treatment to be easier. Different strategies are utilized in solving farm problems.

7.7 MARKETING OF ORGANIC BEEF

Over the years there has been a notable increase in demand for organic beef. This is because there is a growing acceptance of organic products. People are becoming sensitive to have they are eating. People do not want to eat healthier food options free from chemicals. This is as a result of the health effects brought about by eating foods having high levels of chemical composition. The increase in demand has been noted in consumer behaviours. Statistics from different countries indicate a growing demand for organic beef. In countries such as the USA, there are beef products in high demand compared to others. Future predictions state that there will be an increase in demand for such organic products (Lovarelli *et al.*, 2020). The increase in demand for organic beef may be due to the need for natural food stuffs that are highly nutritious. It is known to contribute to good health as it has a high nutritional value.

Figure 7.3: Proper marketing strategies should be put in place to ensure consumers are aware of organic products.

Source: http://www.differencebetween.net/science/nature/difference-between-grass-fed-beef-and-grain-fed-beef/

Organic beef is said to have important proteins needed for good health. Consumer concerns are said to fuel the need for organic beef. This will eventually lead to the global demand for organic beef in global markets. This is noted in market research. Market research shows that millennials are more concerned about their health and are opting to consume natural or organic products. Individuals in this group want products rich in protein and should be a healthier food option. There is substantial increase in demand for organic products. Studies done on organic beef marketing have indicated the following:

Processed organic beef is highly marketable when it is processed. It has been noted that consumers opt to purchases processed beef products. It is predicted that there will be an increase in demand for organic beef. The market for organic beef could also increase to a level that could cause dominance in meat marketing. This could result in the growth of retail stores for selling meat products. Among the most marketable organic beef products is ground beef (Manuelian *et al*., 2020). This is because most beef products are made from ground beef. The development of online retail stores has resulted in the growth of organic beef markets.

In the production of organic beef, the targeted audience is health-conscious consumers. Manufacturers are also targeting foodservice providers such as restaurants. The challenge most face is the fact that most of these foodstuffs are expensive. When compared to other beef products produced from other farming methods, organic beef is more expensive. Proper strategies need to be put in place to help stabilized organic beef prices. This could involve boards handling organic beef production deciding on a fixed price on such commodities. Also, when farmers are aided financially in the course of production, this will enable them to lower their prices while ensuring that they get their profits.

7.8 SUCCESSFUL ORGANIC BEEF PRODUCTION

There are several factors farmers need to take into consideration to enable them to conduct successful beef production. When organic beef production is successful, farmers can get good profits. For farmers starting organic beef farming or those transitioning to beef farming, the first step to ensure they have adequate knowledge on organic beef production. This may require them to consult an agricultural consultant or advisor. They may be required to make certain visits to organic livestock farms to get first hand knowledge and experience on how to run these farms. They also need to read publications on guidelines on running organic farms. Once this is done, the farmer needs to evaluate the organic market. They need to have knowledge of the pricing of beef products and the demand for organic beef. This way they can gauge whether the product is worthwhile or not. They can also evaluate whether organic beef products will yield profits. They also need to know the current market trends. For instance, initially, the demand was high for heifers and steers. Recent data shows that there is a growing demand for organic veal calves. Irish organic beef is expected to have a growing market (Mench, 2008).

Since organic beef production may be quite expensive for farmers, they need to research on available grants given to farmers. These grants have been useful in helping farmers cater to production costs. They also need to be aware of organic farming schemes utilized by organic farmers. They also need to ensure that they maximize the payments given to organic schemes. Organic schemes are having an area-based payment. For this type of organic scheme on-farm or off-farm capital investment schemes are funded by certain programs. In most cases, the Rural Development Program is known to fund most organic farming schemes. With the development of technology, there are many websites giving information on organic farming. they should take a short course on organic farming. This way they can adopt organic farming schemes.

Once they are done, they need to choose a certification body. For farmers to produce and sell organic beef, they need to be certified under a body. This body gives a seal for their products enabling citizens to identify organic beef products that are truly organic. The number of certification bodies may vary depending on the country. The farmer will have to apply for certification from one body. They will have to fill an application form. For farmers who are converting to organic farming, they will have to fill a conservation plan. Both farmers will have to pay a certain fee for the certification. The application once done the farmer may carry out organic farming. For farmers converting to organic farming, once the application is done they will be given a conversion date where conservation activities will commence. They will also be given the duration for conservation. Farmers need to comply with given guidelines and regulations. This is because certification bodies usually inspect to ensure farmers are adhering to the requirements.

Once the farmer has been given the date and period for conversion, the farmer is given a conversion plan. This plan gives an outline of all management activities to be done. It also gives details on the changes to be done on the farm. They include the feeds t be given to animals. They also need to ensure certain analysis on the farm has been done. This is because details on the soil found in that area and the faeces of the animal are needed. The conversion plan entails the housing plan to be used by farmers, the health plan and the feeding plan. These plans can be devised by farmers themselves after consulting organic farmer advisors.

Organic farmers need to have a feeding plan. This entails the type of feeds they will give to their animals. In the case of grass-feeding, farmers need to ensure they have a consistent supply of forage. This is because farm productivity is highly important. To do so they need to ensure the stocking rate is maintained. The forage given to animals should be of high quality. Most farmers opted to introduce white clover to pasture to ensure grass production is maintained. White clover is known to introduce a good amount of nitrogen into the soil. For farmers who cannot access white clover, they are advised to use red clover. The different types of clover when given to animals results in good beef production. This is because they are a good source of protein for animals (Mingmongkolchai & Panbangred, 2018).

There are times where the soil nutrient level is not high enough to meet the demand for forages. In this case, concentrates may be used to supplement the animal diet. Farmers may face the challenge of purchasing such concentrates as they are very expensive. This is why farmers are advised to heavily rely on grass as it is more economical and reliable. They still have the option of using home-grown grains. When home-grown grains are not enough to meet the need, they can purchase organic grains from other farmers. Farmers growing their own feeds are required to use regular toppings. They help improve grass quality. They also need to conduct weeding activities. This prevents the likelihood of animals feeding on weeds.

Animal health is also a factor to consider in organic beef production. For beef from animals to be sold, they need to be healthy. This is why animal health and welfare are covered under organic farming principles. Farmers need to know that drugs such as antibiotics and anthelmintics should not be given to animals. Organic treatments should be given to animals. To find out whether an animal is suffering from any disease a faecal analysis is done to establish what they are suffering from. When managing animal health, farmers need to have an organic farmer's record book. Animal housing also needs to meet the organic farmer's board. The houses need to be spacious enough to allow the animal to move. Beddings should also be provided for the animal. They could be made from conventional straw.

7.9 BENEFITS OF ORGANIC BEEF

Consumption and rearing of organic beef have numerous benefits on consumers, farmers and the environment. With regards to the environment, organic farming has been said to positively influence it. Wastes from animals can be utilized as manure useful in improving soil nutrient and soil fertility. This method of regaining soil fertility is beneficial in that concentration of nutrients is balanced implying that it will not negatively affect living organisms in the soil. In the feeding of organic livestock, farmers need to ensure that there is sustainability of these feeds. Farmers have resulted in planting cover crops. This is beneficial in preventing soil erosion. Cover crops have helped fixed soil particles together make them less vulnerable to agents of soil erosion. Organic farming requires the use of natural and organic products in the rearing of organic livestock (Mie *et al.*, 2017). Most farmers can get these products by planting them in their farms at home. This has positively contributed to reduced environmental pollution. Still, on matters of pollution, organic farming seeks to minimize the use of chemicals in rearing animals. This reduces soil and water pollution.

For consumers, there are beneficiaries of organic farming. Organic beef is highly nutritious and can help improve their health. It is a healthier beef alternative for consumers. This is because the animals are fed on organic feeds allowing them to gain nutrients. Organic meat is a good source of nutrient supplements such as Omega-3. When compared to other beef products produced from other farming systems, organic beef is healthier as it does not lead to health complications. This beef has a fat composition suitable for consumption preventing blocking of blood vessels once consumed. They also have a higher protein concentration compared to other types of beef.

As antibiotics are not given to these animals, it ensures that human beings do not develop a resistance to antibiotics. When human beings consume beef products having antibiotics in them, human immunity develops a resistance to these antibiotics as it perceives them as a threat to the body. If a similar kind of antibiotic is used to treat the person in question, it may not be effective. For farmers, organic farming is a beneficial system used in this kind of production is much cheaper compared to other methods of production. With the growing demand for organic products, farmers rearing organic livestock to get surplus profits from selling their beef. Organic beef has been useful in improving the state of food security. It contributes to global food production. Recent trends in organic beef include the use of technology in its production. This entails technology being used in feeding, breeding and slaughtering organic livestock.

Chapter 8 Organic Sheep

CONTENTS

8.1 INTRODUCTION

When you talk about organic sheep, we put into consideration high standards of animal welfare. They should be bred and raised using specific standards that have been outlined and stated by the relevant bodies. There will be certified agents who will inspect the process from the beginning phase up to the last phase which will be processing the sheep products. The management practices and materials being used should be recorded by the farm to see through that the sheep are organically bred.

For the sheep to be considered organically bred, it should follow all organic managed practices. Such practices involve; Feeding the sheep food free from genetically modified organisms (GMO) and other animal by-products feeds. Pasture grass is a good choice of feed. In the case of lambs, they should be given milk that has come directly from organic sheep; the organically raised sheep should be separated from their counterparts to avoid any form of interbreeding. Hormones used to enhance growth should be avoided and use of sub-therapeutic antibiotics should be avoided at all costs. The use of vitamins and other supplements can be used to improve the health of the livestock and if the sheep gets sick medication can be administered. Vaccines can be used to prevent diseases. Organic sheep should be exposed to outdoor conditions such as pasture, air, sunlight, shade, areas of exercise, etc. Sheep intended for meat products should be produced using organic processes during the last one-third of the gestation period (Miranda-De La Lama *et al.*, 2017).

Sheep as ruminants can harness energy in form of alarm from the sun as they can graze and digest good materials that humans are not able to. Once they have eaten and digested the plants, they can produce high-quality protein in form of mutton which we can digest as human beings. Furthermore, sheep have the ability to utilize marginal land that can not be used for cultivation.

8.2 PRINCIPLES OF ORGANIC PRODUCTION

Below are some of the principles that should be considered when breeding organic sheep;

You should employ renewable resources in your farm and agricultural activity.

In every stage of production, the farmer should be able to maintain the organic integrity and important qualities of the organic products by emphasizing that everything is left organic at every process throughout the production. The health of the sheep and other livestock should be a priority and therefore one should employ alternative care to promote this.

In every stage of production, you should reuse and recycle materials to increase effectiveness.

The biological diversity in the farm should be maintained at all stages of production.

The long-term fertility of the soil should be adhered to by employing organic processes in maintaining this.

Care about the environment by protecting it at all costs by minimizing soil degradation, Soil erosion, avoiding environmental pollution, optimization of biological productivity and enhancing the overall Ecosystem within your jurisdiction.

8.3 RAISING ORGANIC SHEEP

Sheep are normally kept in flocks and can either be put in barns or an open area. They are commonly grazers and therefore they can access food at ground level or low level. For sheep that are put in barns, they are normally fed at least twice a day but most of them are let to just graze around the field as they look for grass and natural vegetation. Sheep need water and there provided for in troughs and containers although in areas that have a lot of moisture the amount of due in the grass they eat is enough to sustain them through the day. In countries such as New Zealand, most of the time there is enough due in the grass to sustain their head in most times. Once the lambs have become of age they become weaned from the mother's milk and are fed on hay, pasture and grains. This process is vital because of the increased competition between the lamb and the ewe for food.

Moderate temperatures are conducive to the survival of sheep. Do not be fooled by the thick wool it processes. Just like humans and other animals, wind chill and heat index factors play a major role in determining the optimum conditions of a sheep. The atmospheric temperature can be different from the body's temperature. During the cold season that is winter, it is vital to check on the cold stress in sheep. One should check on symptoms such as shivering, teeth grinding, huddling, and reduced grazing period times which subsequently increases food intake in the barns. Appropriate facility management practices should be implemented such as; Installation of windbreaks, necessary adjustment of inflow and outflow of air, dry bedding and putting up shelters.

Installation of a heat source equipment in their shelter is also important especially for vulnerable livestock like newly born lambs, freshly shorn sheep, lamb with reduced body temperatures and ill livestock. Farmers are advised not to use heating equipment that is used for poultry (brooder lamp). Rather they should get heat source equipment conducive for sheep. The lamps should be placed high hanging on ceilings to avoid the occurrence of accidents. A shelter will require to be well protected against harsh environmental conditions such as wild predators which might take advantage of vulnerable lambs.

Shelters should have adequate space whether it is temporary or permanent. This is because there should be room for expansion. Fully grown ewes require space of about 13 to 17 square feet each whereas rams should have space of about 16 to 20 square feet each. For the pairs of dam and offspring, an additional 14 to 20 square feet should be provided. Space during feeding should also be considered. This is because large animals might harm or injure small animals when feeding. If there is simultaneous feeding of both large and small animals, then 13 to 21 inches should be allowed for each sheep. Vertical bars should have a width of about six inches (Nalubwama *et al.*, 2011).

A constant and plentiful supply of water should be the number one priority in the shelters, especially during the dry season. In the case of lactating sheep, an average of about 6 gallons should be provided. During the cold season, water should be heated to keep the animals at their required body temperature and to avoid dehydration since they become reluctant to consume water. Placing water troughs along the outside walls should be avoided during cold seasons as it may enhance the freezing of the water. Doors of the shelters should be located in a direction where there will be maximum sun exposure and electricity installation on shelters might be vital to enhance warmth in the shelters. It also helps with enabling proper airflow. The shelter should also be placed at a higher ground to avoid the possibility of flooding.

Adequate ventilation is important in shelters as it enhances the health of the animals. Even during the cold season, there should be proper air ventilation to eliminate dust, gases that are emitted from the decay of manure, excessive moisture and pathogens that might be lingering in the air. This is because the combination of these products can cause lung damages, bronchitis and overall loss and reduction in yield. Few factors can indicate that there is poor ventilation in the shelters which include heavy condensation on walls or ceiling, strong manure odour, animals with runny noses and/or coughing and excessive heat emanating from a shelter. Installation of curtain walls and cupolas and fans can help facilitate ventilation. Wood is the best material to use in constructing shelter. A proper pressure-heated wood will eliminate the possibility of animals chewing on them. Wood is considered to be the best construction material because it provides conditions that can help in controlling temperature. Concrete is the best for flooring although it can be costly and might get cold during the cold season thus it should be adequately bedded.

8.4 NICHE OF THE ORGANIC LAMB MARKET

There is a large market all over the world looking for organic products from sheep ranging from meat, milk, wool, etc. Organic products are already booming in unions such as the European Union and African Union. From 2015 to 2016 the demand for organic sheep products increased by 5 pc. Countries all over the world have more than enough consumer demand for organic sheep products. In Canada, organic sheep production has a very small specialized market. This is because there limiting factors that need to be addressed sooner or later. One of the limiting factors is that in Alberta there is a shortage of certified organic feeds and most sheep products are thus not considered organic leaving available organic products to be limited in supply and very expensive to obtain. Another limiting factor is that for lamb to be deemed organic the lamb meat needs to be handled in an organically certified and regionally inspected meat processing facility. With such constraints, it makes it very expensive to set up businesses that deal with organic sheep as that means that supplements need to be transported from other areas and products cannot be found locally.

In conclusion to set up an organic sheep business in Canada, one need to have a solid business plan and one needs to ask these questions

What and where is the target market of your organic sheep products? How large is the network? What type of mutton cuts does your target market need? If you cut the lamb are you able to sell all of it at once and what are your customers willing to pay for your lamb?

What is your competition? Are there any other businesses providing the same services? What share of the target market do they have control in? What is the time period they have been operating their business? How has their business been thriving?

What is the required quality acceptable in the market? What is the right procedure to follow in every phase of production i.e. production, processing, transportation, marketing and selling? And will your target audience pay enough to sustain your needs?

Is there a certified slaughterhouse within your area? Where is the location of the organic lamb processor? Are they willing to cater to your demand and supply needs? How many lambs can they take? When do they need the lambs? What kind of intelligence does the manager of the meat processor able to follow and record? Can the intelligence gather be given to authorities and customers when needed to? What are the costs?

Are there legit and well-certified organic materials in your location? How readily available are the organic feeds in your area and is there a contract that is needed to be signed? What kinds of feeds are available? Do they match with my needs in terms of quality and quantity? Are they seasonal? Is delivery of the feeds readily available throughout the year or is it available only during the harvest period? Are the quality of feeds testable by relevant bodies? What will be the expenses incurred i.e. buying of feeds, trucking, handling and storing?

8.5 COMPLYING WITH ORGANIC SHEEP PRODUCTION

As shown above, every jurisdiction must comply with the organic production standard that has been set. Just because someone has sheep and rears them with organic pasture and vegetation and does not give them any chemical or artificially manufactured drugs does not mean that the product is deemed to be organic. Strict rules and regulations in all the production phases must be adhered to in order for a lamb to be considered organic. Organic lamb standard of production must be able to have followed the forage-based system, which involves the way you graze your sheep, management of pasture on which the sheep live and the provision of the appropriate non-pasture feeds over the non-grazing winter period. Organic sheep production systems need to develop depending on the location where it is situated as there are factors such as climate and feeds which vary in every area (Papatsiros, 2011). Every organic sheep production must have financial viability for it to function and that is why a feasibility test should be undertaken before venturing into such a business. Every farmer is also required to be responsible for his or her own organic lamb meaning there are no trade-offs.

In order to have a successful lamb management system, there are areas one needs to put focus on at least one year prior before venturing into such a business.

Market- one should be able to identify and understand the market to which they will be selling their organic lamb products that are identification of the target market, the criterion to which the market expects you to deliver your product and the nature of the contract agreement with the market.

Flock nutrition- the farmer should be able to understand the quality and specification of the feeds that the sheep will be taking to optimize the quality of product, access to nutritional advice regarding your feed is also vital and finally food testing and ration balancing.

Lamb health and well-being- Type of care, preventive strategies that will be employed to prevent diseases, approved medication, regular check-ups and advice from the veterinary clinic.

Cost of production management- one should focus on these main factors that are labour and animal feed, nurturing and developing skills necessary to manage organic sheep production and looking for someone who can advise you on matters regarding financing your business.

Facilities and efficient handling systems- a farmer should create a framework that will optimize operations by enhancing efficiency in every production phase and the model should take into account both human and animal welfare for example lambing and lamb feeding.

Fall back plan- one should be able to account for unforeseen circumstances such as drought, floods and diseases.

Figure 8.1: Most countries require at least 365 days period of planning in order to certify that a sheep has an organic standard.

Source: https://www.agric.wa.gov.au/feeding-nutrition/supplementary-feeding-and-feed-budgeting-sheep

8.6 CRITICAL ACTIVITIES

For your lamb to be certified as organic, there are vital activities that must be practised. They involve;

- The feed is given to the sheep should be a flock certified organic feed.
- The standards and requirements needed to manage pasture for the sheep should be organic.
- Any artificial substances or substances that are deemed not to be organic should not be exposed to the lambs in any way.
- Slaughterhouses should have organic standards and need to be certified

Clients or consumers require their quality organic products to be planned, recorded and maintained in terms of the following components.

- In terms of rations of the organic lamb pieces, they are required to be at least thirty percent forage at any particular time.
- Minerals and salts mush have an organic certification.
- Apart from winter, sheep should be allowed to eat pasture.
- In the final phase which is processing and handling, it is required that all activities should have organic standards.
- The distributors of straws for bedding should ensure that it is organically certified.
- The wood used in barns and structures should not be treated with preservatives and they should be certified organic (Pérez *et al*., 2019).
- During transportation, certified organic material (animals, feeds, bedding) will at no time be in contact with non-organic certified materials.
- Any vehicle transporting non-certified organic materials is also required to be thoroughly cleaned in between loads.
- Any party given the duty to either transport or handle certified organic materials must sign a declaration that his or her equipment is up to standard.

8.7 KEEPING CLOSE TABS OF THE ONGOING BUSINESS OPERATIONS

A lamb business will need a well-laid out plan with good records of every financial and business transaction. In the case of the organic lamb business, a more detailed and meticulous laid out record-keeping is very vital. It is important to look for certifying agencies that are strict with their guidelines as it has been witnessed in most places that sub-standard certifications are being issued which is bad for business. All information needed should be collected and recorded accurately and clearly. After recording, good storage of the records will come in handy and finally, the stored information needs to be audited by relevant authorities.

- A certifying agency needs might need the following information to be collected to see if standards have been met.
- A field history of at least three years regarding your acre land involved with all the organic production activities.
- Acreage maps.
- A recording describing the rotation of crops and pastures.
- The identity of all lambs is deemed to be practising organic production processes and all their particulars like health records.
- Records showing all land inputs beginning from farming tools, supplements, feeds and labourers attending to the organic sheep.
- Records of methods used in making manure and compost materials
- Products and procedures employed to control pests, weeds and diseases.
- The system of storage is used to store farm tools and inputs.
- The description and type of input used.

In Canada, under the Canadian Sheep identification program (CSIP), a radio frequency identification tag should be installed in all lambs before they leave the land where they were born. In the process of organic lamb and sheep production, these identification tags will be vital during record keeping. For identification, these tags are normally placed on the right ear whereas the other tags are placed on the left.

8.8 ILLEGAL PRODUCTS AND SUBSTANCES IN ORGANIC LAMB PRODUCTION

Every country has Products permitted to be used in organic sheep production but numerous substances are being used in conventional sheep farming that are greatly prohibited to be used during the organic sheep production process.

There are numerous potential sources of contamination. This can come from neighbouring farms when they are spraying farm chemicals such as pesticides and drift to your organic farm, therefore, leading to contamination. If this occurs, the relevant people should collect and record this information and subsequently any farming practice that will be employed to mitigate these risks from happening should also be clearly recorded. A conventional farm should be separated from an organic farm using buffer zones to prevent such risks from happening. Highways are also deemed to be rather toxic as substances from motor vehicles will contaminate the organic farms significantly (Prunier *et al.*, 2014).

In the event that conventional farmers what to shift and go organic, they should inform other neighbouring farmer's of their intentions. It will be up to their neighbours to consider their sentiments and take the necessary steps in helping their neighbours to have an organic friendly environment. Clear signage can come in handy for the organic farmer. Meat treated with antibiotics automatically loses its organic standards. According to regulations, if a sheep gets seek or injured and needs medical attention, the animal then loses its organic status once it is treated. The well-being of the animals is considered more important than its status of being organic. Once the animal is treated, it should be clearly tagged to indicate it and its information recorded.

8.9 APPLICATION FOR ORGANIC CERTIFICATION

The process of certifying crops and livestock is quite lengthy and it might take more than three years. There must be at least 3 years from the last time any agricultural substance was applied any prohibited substance to make it have an organic standard. From the well-recorded details regarding your organic transition process, there will be an audit in order to certify your worth.

Some supporting documents that might be needed include the record of the purchase of feeds, breeding stock, tests conducted on soil and water.

Record of tools used during the organic production period will also be required. All purchased supplies over the period should also be given. Once the certifying agencies have approved the certificate, then the farmer can go ahead and make other investments.

8.10 CONVERTING A CONVENTIONAL SHEEP TO AN ORGANIC SHEEP

Numerous advanced plotting is employed in the process of converting a conventional sheep to an organic sheep. Before the onset of the third trimester, a non-organic ewe can give birth to offsprings that are organic provided a transition is done before then. For a ewe to be given the organic status it should have the following characteristics;

Figure8.2: New born lambs from conventional ewes can be deemed organic.

Source: https://www.wikiwand.com/en/Sheep_farming

- The ewe should have unique individual sheep identification.
- Have completed relevant records such as identification, breeding and lambing dates.
- Switch to using organic materials such as bedding, feeds and supplements.

- Records of the conversion process of the program such as the date, feed, storage and capacity
- The ewe should not at any time been exposed to any contaminated substance during the transition.
- Should have records of a continuous three-year organic management program on pasture and forage.

Lambs that were born from non-organic sheep can have the organic status provided that they have maintained the organic management program from birth.

8.11 BUILDING AN ORGANIC SHEEP BUSINESS

Managing of Cost

The price of organic materials in grocery stores looks eye-catching as they are mostly affordable in many places and from a business perspective that means the returns will be high. Operations in the farm significantly change when one decides to shift to organic farming from commercial farming. A huge limitation is the strict rules and guidelines which will eventually reduce the quantity of sheep meaning lower returns in organic sheep farming. There is also the limited supply of organic feeds which might be a constraint. This might force the farmer to venture into growing the organic crops by themselves to cater to the demands and in turn make income for themselves. Other costs might be incurred because there will be the need to do cropping meaning additional farm equipment need to be bought and a subsequent introduction of a new management technique (Röös *et al.*, 2018). The introduction of cropping might alter the economies of scale which might have been the sole source of minimizing costs.

With the volatile nature of the sheep market shifting to organic sheep farming might be a challenge as such farmers rely on large-scale selling of conventional sheep within a year. In the event of the existence of a secure organic market, organic sheep farming might come in handy as it would reduce the volatility in the market. Animal feed takes the largest chunk of production resulting in up to 70 percent. Organic feeds and supplements might just push that cost higher. The sale of wool might be used to increase income but it knows that the sale of sheep is the largest revenue producer.

The second-highest cost of production is labour. Every stage in production requires a certain form of labour ranging from breeding, weaning, lambing, weighing, health care, shearing, grading and management of predation. Therefore, the farm manager needs to come up with a model that will optimize the operation and at the same time lead to effective use of resources. Time is also seen as labor because the farm will need to hire professionals and they get paid by the hour.

Other costs that might be incurred include security costs, construction costs, energy costs, processing costs, distribution costs, etc. A study was conducted in France to see how the price premium can help offset the extra costs incurred due to the transition from conventional to organic sheep farming. The study came up with a few assumptions.

- Organic sheep bred over an extensive system required a higher price premium as compared to conventional sheep production.
- Feeds and labourers amounts up to 70 percent of the production cost in a conventional sheep business.
- An intensive and hastened production system is not conducive for organic sheep production.
- In the conventional system, the intensive process is powered by feeding grains to sheep which is prohibited in organic sheep production.
- Irregular cash flow will always be experienced in sheep production while the overhead cost will always be constant therefore one should seek financial advice regarding setting up organic sheep production.

Market Opportunities

In Canada the per capita sheep consumption has increased by more than 10 percent since the early 2000s. However, the consumption market is thin. Only 2 percent of the Canadian population take organic food and most of the 2 percent take organic vegetables based on the perception of quality, health and sustainable food production. Even though organic meat is purchased by a small percentage of the population, there has been a rise in the demand for that small niche. Certified organic sheep producers can take advantage of that market opportunity (Sorge *et al.*, 2019). According to research, only 21 percent of meat consumers in Canada take an organic lamb and this can rise if the quality keeps improving and the consumer pricing of the product gets adjusted. The study showed that retail pricing is the most important factor. It shows that mutton has a higher price compared to other commonly consumed types of meat in the market.

A good organ sheep producer should look for his target market that is willing to pay a premium on lamb meat. Consumers choose organic lamb as their choice of meat because they want to avoid livestock that is fed from chemically sprayed crops and others don't want animals fed from grains. Such consumers understand they will need to pay for more but there is a limit to what they are willing to pay. This will have to force organic farmers to be creative in finding a way of selling their products at a premium that will enable them to keep their businesses afloat.

Planned Marketing

This is the biggest obstacle facing organic sheep farmers because they need to find a viable market willing to consistently buy their product at a price that will cater to expenses and contribute to a margin. If the farmer has access to a large market, direct marketing can be the best move to use. With little knowledge and experience in the marketing of organic lamb, the farmer needs to set up his own price and look for his market. The farmer needs to ask himself about some marketing plans he can employ.

- Who will purchase my product?
- Will I have enough customer bases to cater to my supply?
- Are my customers willing to pay more for my products?
- How can I expand my customer base?
- What part of my products would my customers love to buy for tailoring purposes?
- Is the different taste of organic lamb meat appealing to my customers?
- Can my premium charge be able to cater to my expenses?
- Are there certified organic sheep processors in the region?
- Will the packaging of my organic meat be appealing to customers?
- Is the processor willing to store frozen lamb?
- How much will it cost to process lamb?
- What will I incur in the process of handling, delivering and institutional payments?
- Advertising costs?

8.12 ORGANIC SHEEP PRODUCTION MANAGEMENT

Breeds and Breeding

There are large numbers of different sheep breeds all over and there is necessarily no good breed than the other. In the case of organic lamb production, cross-breeding offers the best quality and the qualities of different breeds is crossed.

Ideally, it is difficult to find two different breeds of organic sheep to cross-breed, therefore, conventional sheep are cross-bred and the best quality breed is then the transition process of converting it to become organic begins. When doing replacement breeding conventional sheep can be used though they cannot be resold as organic sheep if they have been owned for less than a year (Sonoda *et al*., 2018). That is why it is highly emphasized that accurate record keeping be maintained in order to avoid any sort of confusion. Unlike getting organic sheep for cross-breeding is hard, there are numerous places you can get sheep with quality, different genetics. It is recommended to visit different areas to look for the most desirable genes. Biosecurity should be upheld by wearing protective gear such as boots and overalls during the visit.

It is advisable to choose all your sheep for a specific look to avoid the contact and transmission of diseases. In the real world, there is nothing like a disease-free sheep but you can find well-maintained sheep. If you need sheep from replacement breeding you should go to designated farms and them from there rather than going to a feeder pen or a slaughterhouse. It is also important to check on flock dispersal so that you can be able to located sheep of the same age group in order to avoid compromising health security. Avoid auctions as flock dispersal is significant. It is obvious that all sheep were first raised from pasture but you will find that other genes do better in the grass system as compared to others.

Organic lamb producers place emphasize on various types of traits. Quality maternal traits imply that there will be a subsequent quality economic trait. Such maternal traits involve

- Ability to mother
- Longevity and toughness.
- Thriving well under organically managed systems.
- Capabilities of foraging, produce milk and maintain body condition on pasture

- Good health is increased resistance to diseases.

One should also consider buying a ram with a specific trait depending on your preference. You should be able to have in mind what your ewe flock needs to produce. Either you want lambs with quality maternal traits or quality slaughter lamb traits. This of course is dependent on the kind of target market you are aiming for. Your flocks might be producing either wool or milk meaning you might need to diversity your product horizon. Look for a ram that you think will give desirable traits to the offspring. In most cases, birth vigour that is the lamb survival is the most preferable option. A growth rate is also preferred. Rams that will bring offsprings that are easy to herd on pasture will eventually give you good returns as they are less expensive to maintain.

Get information from the person selling you the sheep. Ask for the best quality sheep and be ready to pay extra. Besides, you will be buying quality. Market lambs are best produced by terminal sires as they have good genetics for growth and carcass. Maternal breed sires will give you offspring that have quality products.

Artificial insemination (AI) is an acceptable practice in organic sheep production. It is a bit more expensive but it has the benefit of preventing diseases from coming into your farm from other animals, therefore, maintaining a high quality of your organic sheep.

There is the prohibition of the use of oestrus synchronization and embryo transfer. Because heat detection is impossible with sheep practising artificial insemination without performing synchronization of the ewes heat cycles might lead to reduced lambing rates and increased cost per lamb being born. A sheep that can survive after the grazing period is long overdue is quality for the organic sheep system and the combination of genetics and good management leads to sheep resilience.

Feeding and Nutrition

In Canada forage plays an important role in feeding organic sheep. It is grown and harvested to be used after the grazing season. With an average of 200 days during winter, farmers need to plant forage and store them once harvested. One might face challenges during winter as there might be inconsistencies in supply which might cause starvation of organic sheep. If found it will be more expensive than planting yours which doesn't improve efficiency in your farm business operation (Sossidou *et al.*, 2015). For you to get a constant supply of organic feed you will need to build a rapport with your supplier and that might be disadvantageous to small-scale farmers as they may not fulfil their end of the contract. Delivery and storage will also add costs. Certifying agencies play an important role in helping you find the best source of feeds and supplements tailored for your organic lamb. Take advice from them.

You should sample feeds and perform a test on them in order to find out quality feeds tailored for your organic sheep. With this, you will be able to know the feed nutritional value and make the proper rations for your sheep. Organizations that sell supplements might help you with that. Feeds should have a certification that they are organic. Though there can be exceptions in the event of extreme conditions. Genetically modified organisms (GMO) are prohibited as they are not compatible with the principles of organic production. Supplements should be grown using non-genetically modified organisms (canola, corn and soybeans). Milk from organically certified livestock should be used to feed lambs.

Organic sheep that have been fed solely by grass tend to grow slowly as compared to grain-fed sheep and this can change their taste and quality in terms of tenderness and moisture. Therefore, the farmer should always look for a market base that will appreciate such kind of meat.

Grazing

To get organic certification, records should show that flocks are extensively grazed on pasture and forage. Sheep being ruminants, can easily convert plants into meat, wool and milk. Some studies have been conducted and have connected extensive grazing of organic sheep with profitability. Nursing lambs are usually grazed hand in hand with their mothers until they are weaned. After weaning they are then transferred to paddocks that have a lower quality pasture. It is critical that the nutritional value of pasture is up to standard for the ewes and the young lambs to ensure proper growth and thus tests must be conducted. In addition, good quality pastures means that organic sheep can be sustained during winter and enables them to reach the desired market weight, reduce losses due to illness and demise.

Due to the short farming season in an area like Canada, having quality pasture will enable the farmer to offset the cost incurred from grazing the land. The grazing field should be tested and regularly checked on fertility, organic matter and water. There are techniques employed to extend the grazing season and they include forage stockpiling, swath and bale grazing and rotations between annual and long-term pasture.

Predation

Organic producers should put security systems in place to cope with the threat of predators that people on your lambs. Reports have shown that coyotes cause a lot of trouble in the area of Alberta. Guard dogs should also be used.

- The health of flock.
- In organic farms, health is a great priority and minimizing illnesses are their concern. The general wellbeing of organic sheep should be looked upon.
- Below are some points that farmers should adhere to have healthy flocks.
- Minimize the number of brought in animals.
- Establish tight biosecurity protocols.
- Create a management framework that reduces stress and cares about the welfare of sheep.
- Make sure that feeds have the required nutritional needs in their rations.
- Expectant ewes should be given special treatment and the newborn lambs should be fed with colostrum.

Medication is prohibited and if used, clear records should be noted down. The use of vaccines is also prohibited but in some special cases, it can be allowed for example in the control of clostridia bacteria diseases. Vaccines that are given in the right manner will not harm the carcass of the organic sheep.

- Parasite management in organic sheep.

An organic farm are heavily related to worms. Over the recent years, studies have been carried out regarding the persistent menace of worms noticed in organic sheep and it is a problem that needed to be researched. In the year 2008, a Canadian scholar and grower came up with a practical handbook called ‘LIVING WITH WORMS IN ORGANIC SHEEP’. This part was written by a Canadian organic farmer called Peter Stockdale. During the same year, areas of Canada experienced the wettest seasons with a 40 pc higher precipitation rate than normal. It was discovered that wet conditions increased resistance to eradicating worms. It was realized that wet conditions increased anthelmintic resistance which added a whole other perspective to the challenges faced in parasitic control (Souillard *et al.*, 2019).

To achieve parasite management, a research in Canada was conducted on two organic sheep farmers. One by the name of Atcham had a flock of 150 ewes and 20 does but has been reduced to 60 ewes and 8 does participated in the sheep parasite research program conducted by the University of Guelph in partnership with the organic agricultural centre for Canada. The research found out that his farm had very low internal parasite counts as compared to other farms. The other participant is Peter Smith who is a certified organic sheep and vegetable farmer who currently has 25 organic sheep with the aim of increasing them to 50. Smith does not use grains but purely pasture to feed his sheep and uses hay combined with vegetable cells once in a while. In the research, none of the course facilitators has a background in organic sheep farming and therefore they aim to collect knowledge. A convention was held in February of 2008 and 23 of the participants were sheep and goat farmers but most of them were mainly sheep farmers who mainly dealt with organic sheep but others had conventional sheep.

From the research, there were main points that were highlighted. The challenge of parasite control both affected organic and conventional sheep. However, according to Paul and Achim, it was found out that the use of organic soil, pasture and other feed management strategies there were numerous merits as compared to conventional farming when it came to parasite control in sheep. It was concluded that organic. Sheep practises significantly improved the environment necessary to help control parasite management in sheep. It was also found that it is convenient that the flocks should be pastured in paddocks and should be frequently moved from one to another to avoid the growth of parasites. This should be done especially in wet weather conditions as they facilitate and provide conducive conditions for parasites to thrive. Once the parasites start forming in the intestines, they lay eggs and once excreted from the animal they are left in the paddocks. The eggs then hatch, then the parasites climb on top of the grass and they are consumed by the sheep which they will go back to the intestine which will lead to it being overwhelmed. Depending on the soil, weather and climatic conditions, the livestock should not be returned to the same paddock for weeks or even months to enable the parasites to die due to starvation. Not only will the parasites die due to starvation, but they will also be killed by the heat of the sun and also the pasture around the area will grow back and any manure would have decomposed by then. The overall parasite load around the paddock would have been reduced because its life cycle will have been altered. The health of the soil is also very important, which in turn leads to the overall health improvement of the organic sheep. The health of the soil could be improved by increasing nutrients, which is done things pasturing and adding manure to the soil. Regular soil checks should be conducted to see the level at which the soil is thriving thus determining if organic amendments should be added to the area. Paddocks should be diversified with different varieties of vegetation such as hay, grass and cloves. This is to benefit both the soil and the livestock (Sorge *et al.*, 2019). Some weed should be introduced to the paddock in order to have a benefit with the livestock, as they may provide anti-parasitic effects when ingested by the sheep. Weeds such as chicory and wormwood help draw up micronutrients from the subsoil up to the roots of the pasture. During the harvesting of hay, it should not be cut to a bottom level as it washes away the nutrients from the topsoil.

It was also found out that grain feeding can be difficult for sheep to digest and might alter their pH and in the long run, grain feeding can affect the overall health of sheep, which might subsequently lead to more health complications. Breeding is another factor that might impact parasites in sheep. Peter selectively breeds his sheep to enable his lambs to become robust. Achim culls his sheep but lets them freely breed. Culling is mostly done on sheep that portray undesirable traits or that have chronic conditions. With both techniques, there has been evidence to show there is increased resistance to worm pressure.

When it comes to the housing of sheep, factors such as exposure to sunlight, cold and ventilation were considered and it was found that prolonged stay in shelters might increase the chances of sheep getting parasites. In the shelters, sanitation, proper ventilation and thick bedding are critical. Proper flock management and soil quality improvement generally led to the improvement of the health of the sheep. Even with the presence of parasites, they were not sufficient to lead to its death and the sheep lived perfectly. Several natural dewormers have been permitted to be used in organic sheep and they include garlic tonic, Hoeggers herbal dewormer and diatomaceous earth to mention but a few.

Chapter 9

Organic Goat

CONTENTS

9.1 INTRODUCTION OF GOAT FARMING

Goats are a multi-functional race that is mainly kept for food. Goats are brought up in many parts of the world for meat, milk, wool and leather. Goats can be raised in several ecological zones. For instance, goats had been raised in almost every part of Nepal. And with people of all castes and ethnicities consuming goat meat, goat farming is becoming increasingly widespread. Goat farming requires a low degree of investment relative to other livestock farming due to low demand for hay and lower feed requirements. Shed building can also be inexpensive, and goats can easily be sold when the farmer wants them. One-third of Nepal's total agricultural GDP comes from livestock farming, which makes it a significant contributor to the country's agricultural system. Nepal's fiscal year 2073/74 is forecast at 10,986,114 goats, according to the Department of Livestock Services. Similarly, the estimated volume of goat meat produced is 65,583 tons or 20%. Since goat farming is a low-cost industry, smallholders and women have generated jobs and incomes. Goat farming has also become compensation for alleviating poverty. Food and feed for one cow are easily grown for 5-6 goats, and farmers can easily grow 2-4 goats in small fields (Sutherland *et al.*, 2013).

Figure 9.1: Organic goat.

Source: https://bootsandhooveshomestead.com/10-reasons-goats-awesome/

The following are some of the characteristics of goat farming

- *Production of sales and development:* After five months of puberty, infants are born and the generation of income is ready for eight to ten months of breeding.

- *Worth and benefits of goat farming:* Food supply: 20.3 percent of the world's total meat production accounts for Goat Meat, where demand is rising. Thanks to its low-fat content and high protein content, goat meat is considered nutritious.
- *Restricted investment can be made*: small farmers can start with goats relative to other livestock farmers. Goat farming would provide smallholder farmers with low investment and minimal land use hence the opportunity to generate income at home.
- *PEWA (ownership)*: Daughters and women families are normally given goats and children like PEWA to create a home-based workplace. This ensures that the Pewa must be able to support goat farming on a sustainable basis.
- *Conversion of feed to edible proteins*: goats can be reared for less feed than pigs and chickens. Day forage and feed are consumed by goats that are stored as high-protein meat.
- *Answer needs*: Goats that have helped fulfil the cash requirements can be sold at any time.
- *Return on investment over a short period*: since goats can give birth to more than one child at a time and three times over two years, more returns are created with less investment than other livestock over a shorter period.
- *Milk and wool-producing goats*: we mostly raise beef goats. Goat farming may, on the other hand, be carried out in fair weather and under sufficient control for milk and wool.
- *Fertilizers dependent on compost*: cattle are an important part of farming. Goat farming is an important source of fertilizer for agriculture. Dependence on chemical fertilizers has declined and the capacity for soil production has increased (Sundrum, 2012).
- *Goats are important to rituals and celebrations*: in some rituals goats are considered precious animals, honouring God and Goddesses with offerings of bucks and dollars.

In areas of the Himalayan and Hill Belt regions where there are no roads, Sinhal and Chyangra (mountain goats) are used for transport.

The marketing of goat farming is very critical.

Providing opportunities and boosting self-esteem through sound management for self-employed people, especially young people and currently unemployed in the country.

Improvement of income by conventional methods of goat farming focused on research.

Reduce imports of goats from neighbouring countries in large amounts (crores of rupees).

Most people of various religious sects love chevon and there is a growing demand for goat meat in the area, but only through a sufficient supply of marketing is possible.

9.2 GOAT BREEDS THAT ARE BEING BRED: A CASE STUDY IN NEPAL

9.2.1 Races in the Field

(a) Chyangra; is 2.500 meters above sea level in mountain climatic conditions, made of fur in Pashmina beast. It normally comes to birth only once a year. It weighs between 27 and 30 kilograms on average. A doe and a buck weigh between 35-40 kilograms. The dollars would carry the products about 30% of their weight.

(b) The Sinhal Goats ascend from 1500 to 2500 meters in a mountain range. It is dark, grey, or white. The male goat can weigh upto 42 kg and the females up to 34 kg. The goat has only one infant a year and has around 200 grams of raw fur a year.

(c) The goats of Khari are raised between 300 and 1500 meters above sea level. Khari's are available in a wide range of colours, commonly called hilly cats or AULE. The average weight of the doe is 15-25 kg and the overall weight of the doe is 25-35 kg. These Khari are born, on average, at the age of 16 months, their first child. The stature of the body is smaller and wiser; Khari may be raised by bonding or grazing. Khari goats are the most common local cattle in the hilly areas of Nepal. Due to its temperature, disease resistance and average growth, the goats of Khari are considered useful in the hills of Nepal.

The following are the characteristics of the goats of Khari:

- It will bear three times in two years.
- In comparison to the foreign breeds introduced in Nepal, Khari has a higher degree of disease resistance.
- It can be grown in a traditional cottage.

- No balance to feed on the highways.
- The risk of infant mortality is poor.
- The meat is gorgeous and tiny.

(d) Terai goats: Terai goats are brought up in the Terai belt of Nepal. It will weigh up to 32 kg of dollars and up to 18 kg. It can also be born three times in two years, if well handled. Usually, there are two kids when they are born.

9.2.2 Nepalese Goats of Various Races

Boer

Boer (Boer cattle) originated in South Africa and can be raised in various climates. The Boer breed is more disease-friendly than most mammals and can give birth to 100 percent of twins and three births every two years. Adult Boers weigh more than 100 kilos, with a mixture of local Khari breeds which can exceed 35-40 kilos each year. Boer was established in 2065 BS at the Goat Research Center for Comprehensive Research. Boers are now commercially growing for leading farmers in many parts of Nepal. Boer's maturity is higher than that of other races. In both cases, before and after milking separation, the growth rate for Boer kids varies from 150 to 300 grams per day (Smit *et al*., 2017).

Figure 9.2: Boer goat.

Source: https://en.wikipedia.org/wiki/Boer_goat

Barbari

It has a compact, solid body with short, steady ears, red or brown spots on the white body and does not require food. It is grown for the processing of meat and milk. Adult females weigh 20-25 kg and bucks weigh 25-40 kg. Kids weigh on average 2 kilograms at birth.

Sannen

It's polled and it's white and low on the front thigh. It will take between 150 and 180 days to be born, with an average milk volume of between 600 and 700 litres. Normally, a baby is born once a year. A doe weighs an average of 68 kilos to an average of 70 kilos. Beetal Punjab and Hariyana have characteristics such as small and intermediate bodies, raised nose, flat and collar ore, short and slender tails; the skin is expected to be of good consistency.

Beetal Punjab and Hariyana

They are distinguished by a large and medium body, an elevated body, a bound body, slender and smooth ears, and short tails. Beetal in Punjab and Hariyana are high-quality skin with features such as a wide and medium-sized body, raised nose, back and forth. Sixty to 70 kg in weight, Sannen is renowned for its pure milk and has no horns. It is a Swiss race. There is white and tiny fur on the front. The goat for 180 days, results in average milk output of 180 litres. It usually births once a year, and every birth has a kid. On average, bucks weigh 68 kg and their characteristics include broad and medium-shaped body, raised nose, back and lime horn, flat and hanging ears, short, slender tails. Adult bucks weigh 46 kg in Switzerland. Normally, they weigh up to 3 kg of originated weight at birth, milk, ear to ear, white and short fur (Tallentire *et al.*, 2019). Beetal is thought to have arrived in the Indian states of Jammu and Kashmir and Haryana. Jamunapari

This goat has been imported and has solid body meat from Sirohi, Rajasthan, India. He has strong hair, a thin, pointed nose, flat, long, weak and sharp ears; kids have 2.5 kg weight and are tall at birth. Jamunapari has a large body, long thighs, centrally angled noses, hanging log-like eyes and flat, short horns, and is situated in Jamunapari village, Jamunapri Buck province. Hybrid goats are being weighed in Nepal. The Cross of Jamunapari (Khari and Jamunapari: Barbari cross) Cross of the Boer (Khari and Boer) Cross of the Boer (Khari and Barbari: Khabari) The Sannen Cross (Chari and Sannen) was initially developed in the dry regions of Sirohi, Rajasthan,

India. The slender, fleshy body of Sirohi is well known. Some of its facial characteristics include a pointed fur, a short muzzle, flat ears, long, delicate ears, and sharp horns facing up and down. Jamunapari in a Jamunapari state village is with a weight of 38 kg, long legs and central curled nose (parrot-like nose), hanging long legs and short horns. The Sannen Cross (Khari and Sannen) is a fine beef race from the dry regions of Sirohi, Rajasthan, India and western Nepal. Some of its physical characteristics include rough fur, narrow, narrow nose, flat, high, low ears, and sharp, upward, and horns. The kid weighs 2.75 kilograms at birth and 70 kilograms upon maturity.

Organic Goat Farming is increasing as a viable source of income. There has been a marked rise in the global annual organic production over the last decade. Demand will continue to expand at an annual pace of between 10-30%, according to industry analysts in many countries, and the global organic market will span US$100 billion over the next decade. Organic milk demand increased more than organic beef. Growth in organic milk processing has increased the range of organic, frozen and dairy products available in several countries with high demand. The International Organic Agriculture Federation believes that treatment, biodiversity, fairness and well-being are the core principles of organic goat farming (IFOAM).

Organic farming of goats can increase animal welfare, conserve the environment and reward rural life. There are many barriers to the development of organic goats, especially in the area of intestinal parasites and proper nutritional management. Food science discovery, as well as disease prevention and treatment, will contribute to increased efficacy in the long run. Many laws apply to organic food certification in different countries and there is an increase in the number of regulations. IFOAM is one of the largest organic farming federations in the world. In many countries around the world, IFOAM certification demands that specifications be accepted. These rules act not only as a marketing attempt to preserve but also support the word 'clean,' but are focused on the continued pursuit of environmentally sound, human-conscious and animal-friendly technologies in farming, disease prevention and management, as well as on active organic goat farms. Awareness of organic goat farming from an economic, environmental and animal health perspective will increase your chances of success.

Are the problems of ecology, animal rights, public sustainability and rural economic vitality tackled by organic goat farms? Organic goat farming is increasingly popular with customers as a sustainable alternative to agriculture and rural life. Natural agriculture is a profitable sector. The fiscal, environmental and animal health problems posed by organic goat farming might not be recognized by both farmers and consumers. Novel in the region, food, health and work problems characteristic of organic farming appear to be underestimated, but promotion and development capacities can be overestimated. Treatment and management of organic sheep and goat farms will take several years to complete (Thornton, 2010). Effective organic goat farming requires expertise such as marketing, animal welfare, specialist veterinary advice and technical advice and assistance, as well as determination and perseverance. Consumers and producers typically see animal welfare, environmental protection and crop achievement as key factors in the production and distribution of agricultural products. Requirements should be viewed favorable based on goat milk, beef and organically grown cotton.

9.3 HISTORY OF GOAT FARMING

Near East Neolithic farmers started raising small herds for meat and milk about ten thousand years ago, as well as for fur, bone, muscle and sinew clothing and construction. Goat farming is environmentally sustainable at the moment. Mechanical revolutions (1880–1930) and chemical revolutions (1920–1950) had a major effect on modern, mostly non-organic livestock farming. In the 1970s, research–producer-consumer processes were frequently puzzled by the biological evolution of genetically modified organisms. There are reportedly more than 750 million goats in existence worldwide. You can find goats in most countries of the world, in almost any climate and for a variety of reasons. There are more than 1,000 adapted breeds. About 90% of all goats are found in developing countries and others are called "poor cows."

Rudolf Steiner (1861–1925), one of the founders of organic farming, proposed biodynamic farming based on celestial and telluric influences. Humanity and living objects have been introduced, and from the Greek word bios, biodynamics means life and dynamism, which means power, has been extracted. Biodynamic farming is a creative and natural approach that predates the term organic farming and combines much of the principles of organic farming. This technique is also known as biodynamics. Biodynamic cultivation underscored the individualization of the plant by taking little to no external products. Decreased food safety was blamed on organic farms, synthetic fertilizers and pesticides. The quality of food created was supposed to have been dictated by divine shortcomings, and the world and everything within it were assumed to be both religious and normal in nature-a philosophy known as “monism.” Certification boards were set up in the 1980s. Popularity increased in the 1990s. The term ‘biodynamics’ was coined in 1924. The Rodale Farm was the first modern organic farm to be developed in the United States in 1947. In 2006, it was estimated that the overall turnover in organic farming was USD 40 billion. For the last ten years, it has increased at a rate of 20% each year. Business analysts estimate that demand in many countries will continue to rise by 10–30 percent per year (Vaarst *et al*., 2012). The world organic industry will be worth US$100 billion within the next ten years. According to the Organic Trade Organization, net organic sales, both food and non-food is projected to hit $21.2 billion in 2007, with an expectation of over $25 billion in 2008. According to the Organic Monitor, foreign sales are forecast to reach $38.6 billion in 2008. Organic farming is regulated in 69 countries, with an additional 21 in the workplace. There are currently more than 468 certifiers available in the area. The increase in wild harvests is a big recent increase in organic farming.

9.4 GLOBAL ORGANIC AGRICULTURE

The largest organic region in Australia/Oceania is 12.4 million hectares, the lowest in Africa is 0.4 million hectares. The high demand for organic products in Europe and North America is between 7.4 and 2.2 million hectares of organic land. Organic agricultural development has a comparatively small region in Africa and Asia, creating opportunities for growth and becoming an alternative production mechanism that encourages and decreases agricultural prosperity in these countries. Due to increased demand for export markets, organic farming production in Latin America has increased considerably over the last decade. Organic farming is 32, 28, 24, 13, 2 and 1 % respectively, in Latin America, Europe, Africa, Asia, North America and Oceania. Despite having the highest biological area, Oceania has the lowest number of organic farms. 84% of all farms in Latin America, Europe and Africa combined make up organic producers. With China (2.3), Argentina (2.2), the USA (1.6), Italy (1.1), Uruguay (0.9), Spain (0.9), Brazil (0.8) and the United Kingdom (0.8), Australia is the world’s leading country with more than 12 million hectares (0.6) (Vaarst *et al*., 2012).

The highest proportion (29 percent), followed by Austria (13 percent), Germany (12 percent), Italy (9 percent), Estonia (9 percent), Greece (8 percent), Portugal (7 percent), Sweden (7 percent), Latvia (7 percent) and

Timor-Leste, Liechtenstein, are the largest organic countries (7 percent). (7% of the total figure) The importance of organic livestock production means that more than 60 percent of the organic land used in agriculture in the world is permanent. The remainder includes arable land, barley, cropland and property without evidence of use. Organic cultivation increased almost steadily from less than 7 million hectares in 1998 to more than 30 million hectares in 2006. Organic farming has developed almost continuously. Another significant factor is the field of biodiversity, which grew from almost zero in 1998 to more than 33 million hectares in 2006 and continues to grow. Any land has not been treated as organic as it should be but is eligible as there has been no use of chemical fertilizers or pesticides in the past. This is particularly significant in the cultivation of large-scale livestock, including goats. It helps farmers to switch quickly to organic farming and leverage resources in marginal regions. Organic cultivation in all continents has improved from 2005 to 2006, according to the most recent figures, except in North America.

Over two decades, Europe's organic farmland has increased by 7400 percent, the market is rising rapidly and the demand for organic products is growing. More than half of the land used for productive agriculture in developing countries is permanent grassland, complemented by non-specific (12 percent) cropland, permanent crops (9 percent) and arable land (9 percent). (8 percent of the total). The proportion of organic wild species in Europe (28%), Africa (25%), Asia (24%) and Latin America (24%) is nearly the same all over the world (22 percent). North America is the only continent with a small share of 1.3 percent of the wild. Organic farming is increasingly widespread in Latin America. In Colombia, organic farming increased from 25.000 hectares in 2002 to more than 55.000 hectares in 2007. Organic agricultural exports in Bolivia rose to just under 10,000 tons in 2005 from none in 1995, while total organic production increased from less than 1,000 tons to more than 20,000 tons at the same time. There is a lack of awareness about organic goat growth.

Greece is leading the European Union to grow goat with 37.2% of the total population of goats, followed by Spain (21.6%), France (9.4%), Italy (6.9%) and Romania (6.9 percent). (6.5 percent of the total figure). Although the total number of goats in Europe has been relatively stable or slightly declining over the last few decades, Romania's trend has been rising since 2005. Greece has seen a significant increase in the number of goats used in organic farming, from just over 50,000 in 2002 to nearly 300,000 in 2005. Organic cultivation of goats in Italy and Spain has increased or fluctuated considerably, and since 2001 France has remained largely steady. In organic farming, Italy has the highest number of goats, at about 9% followed by Greece, France and Spain (Ankamah-Yeboah *et al.*, 2019). Martini and Lorenzini have given a list of organic goods derived from dairy and cattle processing systems in the Mediterranean region and the economic value of organic cattle farms in southern Europe. The estimated number of EU-15 cattle is 2.4 million, or 2.4% of the total number of cattle. There are a considerable number of registered sheep in the EU-15 in the United Kingdom and Italy. Half of the licensed goats are from Greece.

9.5 INDUSTRY PRACTICE

The World Organic Agriculture Federation is the world's leading organic farming federation (IFOAM). IFOAM-certified standards are recognized in many countries around the world. These laws not only defend the term "organic" but also encourage it. IFOAM has 750 affiliated organizations in 108 nations, both politically and grassroots. Fitness, sustainability, justice and treatment are the four principles that it covers and encourages the introduction of organic farming systems that are environmentally, socially and economically viable around the world. These principles have been developed to protect, emulate and conserve the biodiversity of soil, animals, animals, humans and the earth as a whole; to promote equity of climate and life chances and to safeguard the health and well-being of current and future generations as well as the environment. These values have been developed for the conservation and preservation of the natural environment. The IFOAM seal is distinguished in various forms: This market-oriented marking uses the goods identified by IFOAM Approved Certifiers to ensure that wholesale suppliers, producers and consumers are certified organically in compliance with the IFOAM Bio Guarantee Scheme. Other seals used for similar variations and purposes around the world are the USA Organic, Australian Certified Organic, INDOCERT in India, Organic Farms in Europe, JAS in Japan, Bio-Gro Organic Certified in New Zealand, BIO in Germany and Bio in China. China has its unique organic method of conversion, which makes it possible for products to become organic over time. It plays an important role and at the same time recognizes its economic value in encouraging organic conversions.

One of the most important aspects of organic goat production is the certification process, which guarantees that certain standards of the sector are required. While the principles of governance tend to be the same, they can change from country to country. A collection of standards and legislation must be met to be accredited organically by the United States Department of Agriculture. Many countries have their organic certification marks, such as the Oregon Tilth Certified Organic Program or the California Certified Organic Producers.

A collection of standards was developed by the Canadian General Quality Board and the Canadian Organic Consulting Board, formed in 1992. EU law is being observed in many European areas in line with the Farmers' Directives of the Ministry of Agriculture, Forestry and Fishery in Japan. There is a national standard for organic goods in China. Mexico's National Organic Products Act shall be enforced by the Ministry of Agriculture. Organic values have also been developed in East Africa by the citizens of East Africa (Åkerfeldt *et al.*, 2020). Certification addresses a variety of objectives, including the fulfilment of increased global demand for organic products, security, and avoidance of fraud and promotion of trade. The method of organic certification varies by region but typically involves: farm facilities and processing methods must comply with legislation which may entail improvements to the plant, supplier of supplies (Compliance); extensive paperwork is required, details of farm history and actual set-up (Study); thorough reporting, farm history and current set-up (Compliance) Annual on-farm reviews, including a physical visit, a record review and an oral interview; an annual fee bill; Daily farm and sales papers, detailing all activities, must be made available at all times for review (Records) (Record-keeping).

The definition of "clean" will be used in pesticides, chemical fertilizers, and all other changes as mother earth do not provide the food we eat today. In Europe, the environmental policy encourages renewable resources and recycling by returning surplus nutrients to the soil. It regulates the production of meat and poultry and stresses the protection of animals and the use of natural food. The use of natural pesticides, herbicides, chemical fertilizers, growth hormones and antibiotics is emphasized in livestock and crop farming, while the use of synthetic pesticides, herbicides, chemical fertilizers, growth hormones, antibiotics and gene alteration is avoided. It supports a series of strategies to conserve habitats and reduce pollution. Food products should have formed organically by at least 95%.

The products shall conform to the requirements of the official inspection scheme. Items can be derived from the manufacturer in a sealed container. The requirements of the USDA National Organic Program are satisfied either in the garden, in the wild or the handling method for the marketing of agricultural goods as organic. It develops, imposes and supervises national standards for organic food processing, storage and labelling, as well as certification agents (both international and domestic) that inspect organic production and management and ensure that they comply with the USDA guidelines.

9.6 ORGANIC GOAT PRODUCTION

All the main aspects of organic goat production include staff, livestock (animal and health care), facilities, management, environmental protection, food quality and safety and profitability. Food and pasture, fertility, breeding, animal health, wildlife protection, environment, plants and vegetation and staff are all essential components of successful organic goat harvesting. Descriptions of organic livestock processing are given for the USDA National Organic Program. In short, organic and non-pesticide pastures, herbicides, chemical fertilizers and all other banned products must be certified. Food served to children must be registered organic (hay, rice, pellets and milk substitutes). Fresh food does not contain synthetic hormones, carbohydrates, coccidiostat, urea and all other limited additives. Organic certification is also required for bedding that can be used by livestock. No animal used for slaughter should be used with antibiotics, anthelmintics, growth implants and other small medicines.

Ivermectin can only be lost based on the volume of faecal cells, but not on a normal or therapeutic basis, and not in the last three-thirds of pregnancy or lactation. Street medicines cannot be handled as herbal animals. Animals are affected by counterfeit drugs. Vaccination is allowed. Up-to-date feed and medical records must be maintained. Animal consciousness is important at all stages of your existence. Human-wildlife, open range and grass-fed animals are not equal to animal wildlife (Alonso *et al.*, 2020). International organic regulation on the use of anthelmintics in the United States is tougher. Internal parasites in confined animals may not be so serious, but they are the worst health problems for goat farmers. Under the European Organic Guidelines, the deworming of small ruminants by specifically selective use of anthelmintics is allowed, as it is difficult to obtain full internal natural control of the parasites and therefore endangers the health of the goat. Usually, organic goat products cost more. Manufacturers must recognize that organic feed is not always readily available and that more land is usually required.

9.7 CARE OF MILK

The production of organic milk goats has been both productive and sustainable. In Germany, Rahman measured the average annual milk yield from a herd of 20 to 35 lactating goats between 532 and 835 g/day over six years. The availability of drill, which had an impact on precipitation and other environmental factors, contributed to fluctuations in milk yields over 6 years. In a more recent study/survey of a larger herd over four years, Rahman later estimated milk yields ranging from 488 to 790 kg/year or between 1.6 and 2.6 kg/day over 305 days of lactation. However, in organic dairy goat farming, higher yields are possible, depending on pasture conditions, according to these reports. Moroni has studied intramammary infections, the development of milk and the consistency of organic dairy chicken, which concludes that, while ensuring potency and efficiency, viruses and somatic counting can be controlled without the use of medications.

9.8 THE PRODUCTION OF ORGANIC GOAT MEAT

There are no significant differences in the production quality or precision of organic and conventional goat herds. In the case of Blanca Andaluza, Guzman analyzed the carcass yields of 24 identical twins from birth to 8.4 kg. There was a distinct index of compactness and offal dispensing, but the range and form of the carcass did not exist. Zurita tested 89 kids from the birth of Murciano-Granadina up to 7 kilograms and observed a difference of 0.5 kilograms between herds. Organic goat meat production typically prefers a food strategy that avoids energy-intensive diets that result in an increased rise in weight. As such, internal fat near the heart and kidneys is most likely to develop. This makes it preferable to have a slower rate of growth and higher market weight.

9.9 MANUFACTURE OF FIBERGLASS

The details on organic fibre goats are incomplete. Using a linear programming model, Eik and Asheim concluded that it could be beneficial to turn herd production into cashmone if the yields of meat and cashmere are improved by feeding children before the age of 20 months (19 kg). They also advocated the achievement of organic fibre production in goats through the advantages of high-value cashmere production, organic weed control, and brush management, and the continuing financing of organic farming (Bessei, 2018).

9.10 ORGANIC GOAT PRODUCTION OF MARGINAL SOIL

A study on 23 organic and conventional farms in Andalusia's marginal lands Mena shows that the typical goat development method is underpinned by organic inputs, efficient pasture maintenance, soil quality and pollution, veterinary care and disease prevention regulations (40 variables in 10 indicators).

Figure 9.3: Fiberglass.

Source: https://cen.acs.org/materials/inorganic-chemistry/s-fiberglass-does-delicate-material/96/i38

9.11 ORGANIC GOAT FARMING DIETARY INFLUENCES

The dietary effect of the parasite in goats has been studied. Protein supplementation has demonstrated improved resistance and/or resilience for grazing goats. This is important, as regulation of parasites and disease prevention can be difficult in the production of the organic goat. Protein supplementation can allow organic goat stocks to develop more efficiently by strengthening their immune systems. The need for energy and protein synchronization to achieve optimum microbial rumen synthesis is essential for the growth of organic cabbage. It could be difficult if the goats were on pasture and relied solely on wheat. Comprehension of foods, eating patterns and seasonal interactions can also help to improve production efficiency. Side metabolites that affect the growth of organic goats are tannins and tannins. The excretion of trichostrongyle in naturally infected goats decreased dramatically when sainfoin was applied to the diet. The number of eggs in goat faeces decreased in tannins. Lespedeza decreased faecal egg count and mortality in goats with high tannin condensation. There is no conclusion on the dehydrated neem; neem and seed extract results in the faecal egg figures. The number of eggs in the faeces of the goats was limited by heather enrichment. Scientific evidence indicates that available heather vegetation could be a reliable choice and a long-term solution to controlling the use of permanent white ryegrass clover in Goat growth systems with gastrointestinal nematode infection.

Standard zeolites have been recently studied to inhibit the number of faecal eggs. Legumes play an important part in organic goat breeding. They can fix nitrogen, improve soil fertility and durability, recycle nutrients, control weeds, disrupt disease and pest life cycles, provide high protein feed for cash crops, foster biodiversity and land productivity and work flexibly. Since the use of chemical compounds in organic goat production is prohibited, it is necessary to find alternatives for disease prevention, control and care. Rahman and Seip have been researched and developed alternative anthelmintics for plants and plant preparations. Blue cumin, black walnut, border tree, mullet, common wormwood, cruciferous, shrimp tree, Eurasian wormwood, fargara, fennel, fennel, black garlic, Gambian mahagon, goosefoot, Indian lilac, Kamala, neem tree, pink root, pumpkin, pyrrh, etc. The following is stated in these plants. Hoste found several bioactive plants for the prevention and elimination of diseases. The legumes and woody plants described included Sainfoin, Sulla, cornicular lotus, lotus pedunculate, dorycnium and chicory (heather, oak, hazel tree, brambles). These plant species can help prevent, manage and treat pests and diseases during the growth of organic goat species. External pathogens and rodents are also managed by grazing practices. Among them are preventive (return to purified pasture without parasite), evasiveness (avoid the challenge of worming from dirty pastures to clean pastures) and dilution (relieving the threat of worm by diluting pasture infectivity) (Buller *et al.*, 2018).

9.12 PRODUCTION OF PRODUCTS

Chemical organic milk contamination is not permitted during fermentation. No synthetic ingredients, including preservatives and flavouring agents, are permitted during the production of organic milk and other milk products. Cheese can be made from fresh milk if it is aged for at least 60 days before it is sold. Fresh cheese production requires pasteurized milk. Organic meat cannot come into contact with non-organic meat during processing, and organic meat and product cannot be produced using synthetic additives. It is important to remember that synthetic internal parasites are not required for the slaughter of cattle. Instead, consideration should be provided to the scheme for faecal testing, killing of heavily infested animals, grazing rotation, manure and vector management and intermediate host defence with allowable materials. There are no organic processors in many parts of the United States and around the world. In certain countries, a processor is available for organic and non-organic products. The need for business in the region must be understood and the processor protected.

9.13 PROMOTIONAL BRAND NAME

Market quality, expense, appearance/freshness, taste and family well-being are the main five reasons for the consumption of food items. More than 40% of customers pointed to value and price as critical factors. Over half of consumers in almost every EU-25 country are concerned about the health of livestock and are thus willing to pay a premium to improve animal welfare. Organic food advertises that it uses laws that do not pollute the environment, that do not use chemicals in the media networks, and that do not harass those involved in the process. Furthermore, cattle are ethically grown and treated rather than overworked, conserved at no cost, are allowed to feed and breed naturally, have good food and have access to fresh food and many even states that there is a strict 'no child labour' policy. The sale of organic goat products, traditional practices, opportunities for demand and access to resources should be taken into account. Producers could stop dealing with the purchaser of milk and refine it themselves in expectation of a rise in agricultural profits (Broom *et al.*, 2013). Various products, such as liquid milk and milk-fed pigs, can boost revenue and financial stability by selling one or more forms of popcorn, tofu, fudge, chicken skins, beef or chicken milk soap. It is necessary to note that organic goats fed food yield an income equal to the cost per unit of profit of grain-fed goats. Organic goat fat can be slightly less concentrated and contain more omega-3 fatty acids, linoleic conjugate acid, beta carotene and vitamin E. Many of these measurements are desirable measures of market consistency that can be highlighted in ads. It should be borne in mind, though, that the grazing of goat meat is more tasteful and oxidizing, reducing shelf life and colour.

9.14 SUSTAINABILITY OF NUTRITIONAL AND ECOLOGICAL DEVELOPMENT

The transition from traditional to organic farming has made food security a concern in terms of food production, availability, stability and usage. Unfortunately, the literature on this subject is incomplete. According to Halberg, organic farming will have a small influence on international food, needs and markets. They proposed that certified organic farming would decrease the demand and increase food prices in Europe and North America. In contrast, uncertified organic farming will improve production, lower prices and increase food security in low-input regions such as sub-Saharan Africa. Organic farming will contribute to global food stability, according to an international meeting, but its willingness to do so depends strongly on the political will. The conference also concluded that organic farming has the potential to mitigate new problems such as climate change through carbon capture, improving the quality of drinking water and protecting it through reducing irrigation needs in organic soils, enhancing agrobiodiversity, improving nutrient adequacy through biodiversity and rural development, etc. According to Bagley, organic farming could satisfy the current food demand of the population thus helping to compensate for the negative effects of traditional farming without extending its land base.

9.15 ANIMAL WELFARE, CLIMATE AND SUSTAINABILITY CONCERNS

Housing, treatment, shipping, killing, disease, destruction, starvation and medical care are all important factors for organic goat farmers to address consumer animal welfare issues. Inorganic processing of goat, animal welfare requirements have not yet been laid down. Lindqvist outlined the findings of the animal welfare projects and pointed out that animal wellbeing and safety should be the norm in all animal husbandry, maybe even more so in organic farming. Tybirk described biodiversity, habitat abundance, size and form, functional integrity for habitats and agroecosystems, and landscape integrity as a core ecological concern for future agriculture. In the future, organic goat farmers should consider this ecological outlook. Ronchi and Nardone described land-related problems as degradation and landscaping, weaning of shared pastures, abandoned margins, low levels of agricultural and livestock integration, land fragmentation, the high dependency of farmers on the feed they purchase, reduced variety and increased specialization in mono-cropping (Brito *et al.*, 2020). As the main areas of emphasis for improving the sustainability of organic small ruminant farming, Nardone has developed breeding, feed management and disease control strategies for small ruminants. It suggested that animal selection should take into account the interaction between animals and the climate, which feed management should aim to incorporate agricultural components into livestock on a spatial and temporal basis, and possible chemoprophylaxis solutions should also be made available for helminth monitoring, such as the use of homoeopathic and enhanced geological medicines.

9.16 FORECAST AND DEVELOPMENTS AROUND THE PLANET

In certain parts of the world, the organic food market is generally agreed to be targeted. In certain cases, the demand for organic food is beyond supply. Due to market rates paid for organic products, investment in organic production will rise, but the net result of the ongoing financial crisis will reverse the trend. To protect the industry, the amount of organic regulations is expected to increase. The disparity between regional supply and demand is expected to continue. The distance between manufacturers and consumers persists. The market's maturity is rising.

9.17 DIFFICULTIES AND PROSPECTS IN ORGANIC GOAT PRODUCTION

Goat production is one of the opportunities that will impact organic farming in the future. On the outskirts of the farm, organically grown from the wild, it seems tempting to graze goats. It is capable of raising the profits of goat farmers in these areas. Global certification of organic products, such as goat cheese, beef and fibre, is also a crucial motivation for the development of the organic industry. While energy and chemical costs are high, sustainable organic goat production usually has a financial advantage. Alternative medicines could be enticing as a result of a ban on limited materials for the treatment of infections and diseases. Emerging health issues, animal welfare and growth constraints; epidemiological surveillance of the major diseases in production; breeding studies on disease resistance and commercial traits; nutrient deficiencies in organic systems, livestock breeding, biological control and use of new plants and plant extracts; and the development of an integrated pest management system are all examples of resetting. The greatest challenge to organic goat production in the future is likely to be animal health and welfare, with a greater emphasis on disease prevention and eradication. To achieve the goals of disease prevention, protection and eradication, monitoring and inspection of natural health products will become more important. In the future, the use of marginal soil for organic goat production in hills and hillside systems is likely to be sufficient. Owing to the heavy use of grazing mineral shortages due to soil characteristics and pasture maintenance practices, organic goat production is a crucial factor. Consumers will also be concerned with preserving organic goat milk, beef and fibre products from fraud and quality assurance.

Chapter

10 Markets of Organic Livestock

CONTENTS

10.1 INTRODUCTION

What comes into the mind of many when they hear of the term "Livestock"? Perhaps it means a group of animals that have spent eternity, from one generation to another, coexisting with the human race in the same environment; but different habitats, with the habitats being able to accommodate either species. However, the explanation which sounds reasonable is that livestock was part of wild animals that with time, from one revolutionary age to another, were slowly incorporated into the human environment and became part of them. This belief is reasonable because, it is believed that in the first revolutionary age, the human race was still evolving and that they lived in caves; which are environments that are not fit for accommodation of more than one species of animals (Cambra-López *et al*., 2010). With this, it is believed that as the human race continued to evolve, so did their relationship with wild animals improve, which led to some of the wild animals becoming domestic animals, and being able to reside alongside human beings.

10.2 INTRODUCTION TO ORGANIC LIVESTOCK PRODUCTION

As some of the wild animals turned into domestic animals by being able to coexist with the human beings in the same environment; ways had to be invented to ensure that the animals become fully domestic animals and they do not turn out to be wild animals. Such ways included the feeding habits, their interaction with the human race through; the ways they were being taken care of, the benefits they provided to the human race, among others.

When it came to taking care of domestic livestock, traditional ways were used. One, the animals would be looked after by people in forest lands which were rich in all sorts of livestock feeds and later taken back home in the evening. The livestock would be watered in nearby rivers.

As for the treatment of domestic livestock, at first, they were not treated. The previously wild domestic animals would naturally heal from illnesses and had natural ways of being resistant to disease-causing microorganisms. However, as time went by, the traditional herbalist who only specialized in healing human beings, thought of ways of using their healing methods on livestock, which did work.

Originally, the human race rarely harnessed the products obtained from livestock. But with time, as humans evolved, they developed the desire of having different meal types which led to them discovering the different by-products harnessed from livestock. Back then, the products were harnessed in their original quality and normal quantities.

Basically, it could be said that the original domestic livestock animals solely depended on natural ways of survival. From feeding on natural pasture to being treated using natural herbs and produce from the livestock being obtained in its natural quantities and quality, with no methods being used in boosting the quality and the quantity.

With the advancements in the revolutionary ages, the ways of rearing domestic animals changed. The human race had understood the importance of by-products harnessed from livestock. With such discovery, they sought ways to increase the quantity of the produce and its quality. The methods involved the use of natural components which with their scarcity due to climate change and such like things; artificial components were manufactured to boost the quality and quantity of the by-products from domestic Livestock.

Fortunately, the quantity and quality of the by-products increased with the introduction of a new invention each day. The demand for livestock product is increased, so did the supply and the need to come up with more ways of ensuring that the supply is equivalent to the demand.

Unfortunately, with the increased consumption of the improved livestock by-products, the consequences came into play. The environment degraded due to the use of excess artificial inputs to boost the quality and quantity of Livestock feeds to improve the health and boost the productivity levels of livestock; the health of consumers degraded due to consumption of high levels of artificial components indirectly through the livestock produce; the health of livestock degraded too because their body system had mutated and became resistant to the artificially manufactured drugs; among other negative effects.

With the increase in the above-mentioned effects of using artificial components in livestock rearing, the human race has considered going back to the traditional ways of rearing domestic livestock. However, the journey back is not easy, for the traditional methods of rearing domestic livestock were cumbersome, unlike the current artificial ways which are mostly mechanized. Furthermore, the demand for livestock products is still high, so going back to traditional ways means that ways have to be sought to ensure that the supply for livestock produce meets the demand. This means that the traditional ways of livestock rearing have to be improved, organically of course, to be at the same production level as it was with artificial livestock rearing.

Rearing organic livestock to ensure that their production level is the same as that of the non-organic livestock requires careful attention and that certain norms be practised without questioning.

One, the young ones of organic livestock should be given ample time to grow next to their mothers for as long as possible. This ensures that their immune system is developed right at birth and reduces the risk of health problems on the young ones later in their maturity levels. This might contradict the mission of maintaining a certain production level, but this is where the quality of organic feeds comes into play.

Organic Livestock ought to be fed on purely organic feeds which can boost their health and their production levels. Well-fed organic livestock ends up producing products of high quality and maximum quantity. For instance, a cow that is breastfeeding its calf can produce surplus amounts of milk that is enough for the calf's growth process and also in the right amounts as required by the farmer.

Before organic livestock are incorporated into a farmland, a farmer ought to ensure that the livestock is of the right species. The right species of livestock ensures that it can be termed as an organic livestock and manage to thrive well without having to face any side effects (Wilhelmsson *et al.*, 2019).

Organic livestock should not be exposed to any sort of stress. If it is feeding, it should be readily available in its natural form. As for the water, clean and healthy water should be used into ensure that they do not have to suffer any gut illnesses damaging their quality and quantity. However, this does not mean that the livestock should be kept in one place. They should be allowed to move as they graze, to move as they to quench their thirst in clean water sources; for such movements strengthen their muscles improving their health in a way.

As for the grazing methods, they should be well managed. Open grazing, as stated earlier, allows organic Livestock to stretch up their muscles, but it does also have its own disadvantages. For instance, a vast open grazing field might be hard to manage through the elimination of disease-causing microorganisms such as parasite larvae, which end up contaminating the pasture. Contaminated pasture causes illnesses to Organic Livestock. To avoid such instances, open grazing of Organic Livestock should be done in a medium-sized field, which is easy to manage. By managing it means that, it is frequently cleared to allow for the growth of fresh pastures; it should be cleaned from contaminants such as the parasite larvae using organic techniques such as the use of natural herbs to avoid causing diseases to livestock.

The housing conditions for organic livestock should be top-notch. If it is their shed, the shed ought to be cleaned daily of their waste. They should be fumigated, using herbs, from time to time to ensure that disease-causing microorganisms cannot survive. They should be constructed in such a way that the livestock can have proper ventilation that allows for efficient circulation of air into and out of the shade.

It is advisable that the rearing of Organic Livestock be practised more in that the livestock should be reared in large quantities. However, if the rearing of the livestock is in large quantities, there should be enough pieces of land to accumulate all of them. They should not be overcrowded in one piece of land such that they have to struggle for clean air while in their sheds or struggle for feeds while grazing in open fields. As stated earlier, they should be in a stress-free growth environment.

10.3 IMPORTANCE OF SWITCHING TO ORGANIC LIVESTOCK PRODUCTION

As it has been mentioned in the above section, the rearing of livestock using Organic methods is so beneficial not only to the livestock but also to the human race and all other living organisms which rely on livestock either directly or indirectly.

Organic livestock promotes nutrient cycling in the soil through the use of their waste. Manure from organic livestock is believed to be rich in nutrients and its use in the farm ensures that the depleted nutrients in the soil are replenished. This is achieved through the spreading of the manure on the farm and its mixing with the soil.

As stated earlier, organic livestock is reared using organically produced livestock feeds. The organic livestock feeds help in improving the soil structure for they are intercropped with other farm crops and also from one planting season to another they are rotated in the different planting plots in a farm. The intercropping and the frequent rotations ensure that the soil structure is improved in that; the rate at which soil nutrients get depleted is reduced (Castellini *et al.*, 2012). Intercropping also promotes the diversification of agricultural systems, which ensures that at least one crop type manages to survive despite fluctuating conditions such as a change in weather patterns from one planting season to another among other conditions. In large farmlands, instead of having to incur money in weed clearing, Organic livestock could be of much help. Since organic livestock feed purely on natural feeds or feeds that have been organically produced; they could be grazed in farmlands that have been infested by weeds (bearing in mind weeds are naturally grown green vegetation), and this does eliminate the need for use of herbicides or excessive labour in the clearing of weeds.

Organic livestock is quite strong for they are healthy because of the feeds they feed on. This means that they could be used in preparing a farmland for plantation purposes. For instance, organic bulls could be used to plough the land to prepare it for transplanting crops from the nursery to the field. Organic livestock could also be used for transportation purposes. For instance, organic horses or donkeys could be used to pull a cart transporting different kinds of goods from one place to another, which saves a farmer's money which would have been used in acquiring transportation means.

The rearing of organic livestock does help in giving the land some break from farming activities. The land is left bare, without any farming being done on it and natural vegetation allowed to grow. This natural vegetation is used to feed organic livestock. Such a practice ensures that the insects, pests and disease cycle in farmland have been eliminated. This is because, the chances of naturally grown vegetation being infested by insects and pests and also being affected by disease-causing microorganisms, which previously had affected the grown crops are low. By the time the land is used once again for planting purposes, the cycle of crop destructors will have been terminated and the newly planted crops will be able to thrive well with minimal interference to their growth phases.

Apart from the above-mentioned importance of switching to organic livestock, there are so many other benefits that can be harnessed from organic livestock; from their products to improving health and biodiversity among others.

10.4 INTRODUCTION TO MARKETING OF ORGANIC LIVESTOCK AND THEIR PRODUCE

Due to the increased benefits of organic livestock, their demand has increased. This means that they are being sold in large quantities from the young Organic Livestock to the fully grown organic livestock. Still, the products produced from organic livestock are being believed to be of high-quality hence having an increased demand for them to be supplied to various regions.

For the marketing sector of organic Livestock and their products to survive, the rate of production should go hand in hand with the rate of their supply to the market. It would not make sense to have well-established supply platforms with no organic livestock to supply to buyers or enough byproducts to supply to the consumer. It is reasonable that, the supply chains be maintained based on the production levels.

The relation between the production of Organic Livestock and their produce with the supply chains is governed by a few factors.

One of the factors is that the definition of the end-consumer. Before organic livestock have been marketed, the reasons behind their purchase by a consumer should be determined. It means, the demand for organic livestock is high and it would not be reasonable to sell them to a buyer who will only end up not taking proper care of them leading to the degrading of their quality. If the buyer is a certified organic livestock farmer, at least with that there is the assurance that the organic livestock will be appropriately taken care of. The other issue is with the marketing of organic livestock products.

The supply chains should be specific to the supply of certain products. If it is organic meat, the supply chain should solely focus on obtaining organic meat from several slaughterhouses that do also specialize in the slaughtering or purely organic livestock.

The location of the buyers ought to be determined to ensure that if it is the organic livestock or their products that are being sold; they are transported in the most appropriate way ensuring that they arrive at their destination in perfect conditions.

For organic livestock produce, the state at which it is to be delivered to the consumer should be identified early in advance. For instance, a product such as milk; will be transported in its original form as liquids or should it be processed to produce different by-products before it is transported to its required destination.

The second factor is the authentication of the supply chains. We live in a world where corruption has become part of our daily lives to the extent that you have to bribe someone for you to receive even the most basic services. To eliminate the disease of corruption in the marketing of organic livestock and their produce, the supply pathways ought to be certified (Cornish *et al.*, 2016). A certified supply pathway is at least a guarantee to the consumer that the ordered products will be delivered in their right quantities and quality. The farms from which organic livestock and their produce ought to be certified to ensure that, the livestock being sold is purely organic in that it has been reared using organic methods; and that the produce obtained from such farms is purely organic and that there are no additives that have been used to improve the quality of the produce.

The third factor to be considered in the marketing of Organic Livestock is communication. It has always been said that communication is the rock foundation of any successful partnership. Trustworthy communication channels between the producers to the supply chains to the consumer ought to be established. Quality documentation procedures ought to also exist to ensure that the marketing processes are verified at each stage. For instance, to verify that the livestock is purely organic, a well-formulated report should be presented to prove that the animals have been reared using purely organic methods. A document should also be provided to prove that, the produce from the organic livestock was obtained using organic techniques and that no artificial additives have been added to the products to boost its quality or quantity. With documentation, marketing logos are easily set up and the products to be sold or even the organic livestock to be sold well recorded for transparency.

In recent years, the marketing of organic livestock production has been facing a serious challenge of substitution. With substitution, non-organic livestock producers are being substituted and sold as organic farm produce which affects the trust levels of consumers on organic produce, which is believed to be rich in nutrients and not likely to cause illnesses since they have no artificial chemicals in them. To eliminate substitution, the factor of labelling comes into play. The best labelling techniques ought to be employed in the labelling and the naming of organic livestock producers to ensure that non-organic livestock produce cannot be substituted for organic produce.

The methods used in the production of organic livestock do vary from one farm to another; however, they still stick to the same principle that; despite the type of methods used, they ought to be purely organic. When it comes to marketing organic livestock, a farmer should best market their organic livestock or the produce from the organic livestock based on their farming techniques. The best techniques will ensure that the best of the organic livestock and their produce gets the first slot in the markets.

As stated earlier, some of the organic livestock produce ought to be processed before being marketed. This means that there should exist well-defined processing guidelines. The guidelines are there to ensure that the processing techniques used do not alter the quality of organic produce by adding artificial additives. If the quality ought to be changed, it should be done using organic methods of processing and not artificial methods of processing. The guidelines should exist at each processing level; from transportation in that, the means of transport should not alter the organic produce in any way; to the actual processing; packaging and resell of the processed products to the retailers (Clonan *et al.*, 2015).

Before marketing, the produce from organic livestock should already have been decided whether it will be sold in its original form or there will be the addition of value to it. With the addition of value, it does not mean processing. The addition of value incorporates the mixing of the produce with other organic produce to form a completely new organic product. For instance, if it is organic meat, it could be cooked using organically produced ingredients. The end product of the further processes Organic Livestock produce should be able to prove that it is purely made from organic produce. This could be done through the issuing of a certificate that authorizes further processing of organic produce still using organic ingredients and organic methods of processing. If the further processes organic produce is to be packaged, the packages should be able to at least prove that the product is purely Organic.

10.5 THE MANAGEMENT OF ORGANIC LIVESTOCK SUPPLY CHAINS

Apart from the above-mentioned factors which are beliefs to govern the marketing of organic livestock and their produce; the supply chains ought to be properly managed to ensure that the entire marketing process thrives smoothly.

The proper management of supply chains; be it supply chains for the sale or organic farm produce or organic livestock; is governed by four principles.

One of the principles is organic credibility. With credibility, it means that the producer ought to be able to verify this product is truly organic or this animal is organic. Credibility cannot be achieved only by the producer; meaning that the producer, the supplier and the customer must come together and agree on the most appropriate way that would prove the credibility of a product. They could either agree that both the producer and supplier should have individual certificates that will be stamped upon presentation of the product to a certifiying expert (who is an expert in organic methods of farming). The certifier's main job is to verify that the product being sold by the producer to the supplier is purely organic and also that the produce is supplied by the supplier to the customers is also purely organic. The standard of a certifier should be well-known and with a good reputation. This boosts the belief that a customer will have in the supplier and also the belief a supplier will have in the producer (Clark *et al.*, 2016).

The second principle is documentation. The importance of documentation has been discussed earlier but we will touch on it briefly but from a slightly different perspective. In supply chains, documentation is majorly used for traceability purposes. Traceability means that, if it is the customer, he or she can trace where the supplier got the organic produce from and also the credibility of the producer. The credibility of the producer when it comes to documentation means that the customer can see, in the document, that the producer uses certain organic methods of production. Basically, with traceability, the producer should be able to produce a certificate that allows them to produce organic livestock, the supplier should be able to produce a certificate that verifies they are experts in outsourcing of quality organic livestock produce, and the path back to the farm should be traceable. However, documents can be faked, the products should be tested from time to time to verify that they are purely organic. Basically, documentation is an important principle for it acts as an alert to all the levels in the supplier chain, alerting on the status of organic produce or organic livestock.

The third principle is in co-operations among the key members of a supply chain, which are the certifiers, the producer, the supplier and the customer. With maximum co-operation among the members, the year-round supply of quality products is determined. Year around supply ensures that each member knows at which time of the year they will be playing their part. A certifier will know which time of the year they will be busy certifying organic produce to both the producer and the supplier. The producer will be well aware of the time they are expected to sell their produce to the supplier; the supplier will be aware of the time they are to get the produce from the producer, and the customer will be aware of what is the perfect time they should pick their deliveries from the supplier. With year-round supply, calendars could be created for only that purpose. The software could also be invented to help in updating all the members of the supply chain of the specific stage according to time. Still, the year-round supply could further strengthen the bond between the members of the supply chain. For instance, the supplier will be able to identify that at this time of the year, the producer needs organic feeds. The supplier will then go ahead to outsource quality organic feeds for the producer to boost the production rate and quality of the end product.

The fourth principle involves the selection of a processor. A processor in the chain of supply of organic livestock and their produce can add value to the product or boost its quality in organic ways. Since the organic produce needs to retain its organic components, even after being processed, the processor ought to be certified. The certification should be assigned after it has been proved that he or she knows all the organic methods which could be used in boosting the quality of organic produce; or if it is the addition on value, the processor knows which organic components to use to add value to the organic livestock produce.

In summary, for the entire marketing process to be a success; the connection between the members of the supply chain ought to be well established; the verification of the quality of produce, that it is surely organic livestock produce ought to be automated and made obvious before the selling of the produce; and there should exists credibility in all levels of the supply chain.

10.6 MARKETING OF ORGANIC MEAT

The cost of producing organic livestock is considered to be much higher as compared to the production of conventional livestock. This means that the pricing of organic meat will be high as compared to the pricing of conventional meat.

Figure 10.1: An image illustrating the packaging of organic meat.

Source: https://cookeatpaleo.com/buying-beef-organic-grass-fed-meat-delivery/

Organic meat is basically the meat produced in organic farms which feed their livestock in organic feeds, treat their ailments using herbal medicine and treat their livestock with a lot of care. Organic meat can be from organic poultry, organic chickens, organic pigs, and any organically reared animals or birds.

For instance, organic bulls require high intakes of organic food for them to grow healthy and strong. This means that, for a farmer to reduce the cost of production, he or she has to locally grow the organic feeds within the farm.

However, despite the high cost of organic beef, some individuals prefer it over conventional beef. This is because organic beef is considered to be rich in nutrients that improve the consumers' health. This is because organic bulls are said to have been raised without the incorporation of artificial hormones into their system, without the use of artificial antibiotics which prevents them from being resistant to diseases and that they are fed on quality organic feeds.

For organic meat to be of high quality, more so organic beef, it is believed that certain practices ought to be carried out. One, the organic bulls should be treated with a lot of care and minimal exposure to stress; meaning practices such as the likes of tethering should be discouraged. The organic bulls should have enough space to feed on allowing having free movements around the feeding areas (Chander *et al*., 2011).

As for the feeds, as stated earlier, they ought to be purely organic. Medicines should be herbal and genetically engineered vaccines are strictly prohibited among organic bulls. Growth hormones should never be used to boost the growth of bulls. Furthermore, it is believed that in organic bulls, the longer they take to reach maturity level, the stronger and healthier they will be.

Organic beef is segmented into its own space in retail shops, supermarkets and even butcheries. At the beef segments, organic beef is placed separately from conventional beef. Organic beef can also be placed in retail shops depending on its source of origin; for instance, the region it came from.

Organic meat from chickens, pigs which is the pork, sheep which is the mutton and fish is also highly valued. The rearing of organic chicken, organic sheep and goats, organic fish fillets strictly adheres to the rules of organic methods of farming, which have severally been mentioned in the above sections.

Organic meat is believed to have approximately fifty percent of fatty acids and it has lower fat levels as compared to conventional meat. This means that the chances of organic meat contributing to health issues such as obesity, which later lead to diseases such as heart attacks, are low.

Due to the increased demand for organic meat because of its numerous health benefits, many governments have set aside incentives to help in its production and also its marketing. As for the production, organic feeds are easily availed to organic farmers who find it hard to access such feeds. This ensures that organic livestock is well fed to be healthy and strong.

Government trade organizations have been formed to help in the marketing of organic Meat. This has gone to the extent that some governments have sought foreign markets for organic meat produced locally in their country. As for further processing, governments have gone ahead and supported the setting up of processing factories in easily accessible locations to the organic livestock farmers. This has made it easy for them to transport their livestock to the factories for slaughtering and further processing such as packaging. Since the processing of organic meat ought to be processed in such a way that it does not lose its organic components; special methods ought to be used in the processing. With the help of governments, skilled personnel in the processing of organic meat are employed.

The marketing of organic meat does follow the principles discussed earlier. One, the packages should have a unique stamp that verifies that the meat is purely organic. As for the butcheries which do not sell organic meat in packages, they should have a certificate that proves that they are certified suppliers of organic meat to the consumers. There should also be traceability of organic meat. If it is packaged, further information should be provided on the packages to enlighten the consumers on where the meat came from; this improves the trust of consumers in organic meat for there have been cases of substitution. Further information allows consumers to verify that the organic livestock has been produced using organic methods for there have been cases of mistreatments of organic livestock, with some not being provided with enough feeds for their growth; and later sold in poor health conditions to the consumers.

The profits amassed from the sale of organic meat are said to be rising at a high rate as compared to the profits of conventional meat. This is because, as stated earlier, the pricing of organic meat is a bit high as compared to conventional meat but this does not stop consumers for they do know the value of organic meat. As the saying goes, cheap is expensive; cheap conventional meat results in health issues that are expensive to manage whereas expensive organic meat continues to boost the health of the consumer. The high profits from the marketing of organic meat do help in improving the organic methods of livestock production for they are said to be quite expensive to accomplish. The profits do also boost the living standards of the farmers.

To boost the purchase of organic meat; retailers are advised to clearly label the shelves which feature purely organic meat. There should also be frequent advertisements of organic meat to the consumers as they shop in supermarkets and retail shops and even butcheries. The public should also be educated on the importance of feeding on organic meat: the health benefits among other benefits (Celi *et al.*, 2017). The education forums could be conducted on different platforms such as social media platforms, and even billboards.

10.7 MARKETING OF ORGANIC DAIRY PRODUCTS

Organic dairy products are products that are produced from livestock that have been reared using organic methods of farming. Organic dairy cattle, are fed on purely organic feeds. They are treated using herbal medicines obtained from organic plants. They are not exposed to stress factors such as dehorning or tethering. They are treated with a lot of care. The use of growth hormones or artificial feed supplements is strictly discouraged.

Figure 10.2: An image illustrating the packaging of organic dairy products.

Source: https://www.pinterest.com/pin/395120567313648549/

For a farmer to rear organic dairy cattle, he or she should be certified and well educated on the organic methods of rearing organic livestock. The methods of obtaining organic dairy products should be purely organic.

For instance, the milking process should be done naturally, and if there is a need for mechanization; the mechanization techniques used should strictly adhere to the rules of organic methods of farming in that they will not lead to the pollution of the environment.

Organic dairy products are known to be rich in nutrients. They are advantageous in that they do not require preservatives for them to last long; their fat content is low as compared to the fat content of conventional daily products; they are rich in nutrients. When it comes to the consumer's health, organic dairy products are preferred for they are known to boost the metabolism process in a human being body; they boost the body's immunity by making it more resistant to disease-causing microorganisms; they strengthen bone muscles by reducing joint pain; due to their low-fat content, they reduce the risk of heart attacks to the human beings; and since they are naturally produced, without the use of any artificial additives, they reduce the chances of cancer which is mostly brought about by too much intake of artificial components.

With the above-mentioned advantages, the demand for organic dairy products is slowly increasing. Still, the effects which have been caused due to the consumption of conventional non-organic dairy products have pushed many people to look for ways to reduce them. Such effects, which are the complete opposite of the advantages of organic dairy products include: the rise in the number of heart attack cases due to the consumption of milk with high-fat content; increase in the number of cancer cases due to consumption of dairy products rich in artificial additives end up increasing the number of cancer cells in one's body; low nutrients levels in the body due to the intake of dairy products, which add low nutrients to the body; among other effects.

Most governments have taken the initiative to boost the production of organic dairy products; with the increased consumption of organic dairy products, the congestion in hospitals due to the increased number of patients suffering from cancer and heart-related diseases will reduce. Such initiatives include the funding for educative programs. The educative programs are aimed at educating the public on the importance of consuming organic dairy products. Through education, organic farmers do also get a chance of enhancing their skills of rearing organic dairy cattle for they are taught about the latest organic techniques such as the best herbal medicines for certain ailments in organic livestock. To reduce the cost of organic dairy products, most governments have taken the initiative of supporting organic dairy cattle farmers with organic feeds which save on their production cost. The farmers are also supplied with naturally made herbal medicines by governments saving on the time the farmers use to look for certain medicinal plants, make the concoctions and use them to treat the animals. Mechanization techniques that stick to the principles of organic farming are being taught to organic dairy farmers to reduce the labour used in the rearing of organic dairy cattle.

Governments have also played a role in overcoming the challenge of substitution when it comes to the marketing of organic dairy products. This has been achieved through the breakdown of the certificate assigning process into simple and easy to complete steps. With that, organic farmers, processors and suppliers can obtain their certification in an easy way (D'Silva *et al.*, 2017). This allows them to be able to verify that their products are purely organic and the customer can verify that and also trace the source of the dairy products.

As for the processing, governments have played a part in ensuring that the processing techniques maintain the quality of organic products. That is, no artificial additives are added during the addition of value to the original organic dairy product.

The marketing strategies have been advanced through the support of innovation of applications that specialize in the selling of purely organic dairy products. Such applications make it easy for a large population to gain access to organic dairy products.

Organic dairy products are further categorized into different categories; however, they are all from the same product, which is milk. The first category is that of the fluid milk, which is known to be the original dairy product obtained from organic dairy cattle; the second category is that of Yoghurt which is more of fermented milk which has organic food flavours added into it; the third category is the cheese, which is then obtained from the fat contents of the milk; the fourth category is butter; the fifth is cream; among others. The cheese, butter and cream are obtained from the fat contents in the milk.

When it comes to the marketing of organic dairy products; they are marketed individually, on different shelves, based on their category. As for the packaging, some of the products are packaged into pouches, others in tetra-packs, other in bottles, others in cans, among other packaging types. Most of the dairy products, more so milk, is packaged mostly in tetra-packs for they are believed to preserve the nutritional value of the organic dairy milk due to the tetrahedron shape.

Organic dairy products are distributed into different market channels. They can be found in supermarkets, hypermarkets, discount stores, convenience and grocery stores, stores that specialize in the distribution of organic products, online marketing platforms, and direct sales that are directly from the farmer or the processor among other distribution channels.

10.8 MARKETING OF ORGANIC HONEY

Bees can be produced using organic methods of farming. Actually, back in the days, during the earliest revolutionary ages, bees were reared naturally. All a farmer had to do was making the beehive and hang it on a tree in the forest. During the time of harvesting, they would go harvest the honey, sieve the liquid part and package it in its original form and sell it.

Figure 10.3: An image showing a package for Organic Honey.

Source: https://www.foodpackaginglabels.net/honey-labels/

With conventional methods of farming, the honey being sold is no longer natural. Artificial food flavours, unnecessary preservatives and food colourings are being added to the honey. The additives lower the nutritional value of the honey and in turn, contribute to health issues for human beings.

Due to the negative effects which have been noticed because of the consumption of conventional honey, most individuals are slowly going back to the naturally produced honey products; that is, honey produced using organic methods of farming.

With organic methods of beekeeping, all a farmer has to do is place the beehives in strategic places which are a sure bet that they will organically attract bees into the beehive. During the harvest season, organic techniques which guarantee that the bees will not be harmed are used. Upon harvesting, organic beekeepers use organic methods to separate the liquid honey from the wax. Liquid honey is then packaged in its original form.

As for the processing, no additional value is added to the honey; not even the preservatives since organic honey is known to not be perishable hence it can last for long. At the processing level, organic honey is basically packaged into cans in its original form. The wax could be processed further and used to form other products such as shoe polish, hair wax among others (Dippel *et al.*, 2014).

As for marketing, governments do play a role in ensuring that the farmers are educated on the latest organic methods of beekeeping. The public is also educated on the importance of consuming organic honey to conventional honey.

The pricing of organic honey in the market is standardized. Just like any other organic product, organic honey ought to be verified that-is organic through the stamping of the packages.

Apart from the use of organic honey for consumption purposes, it could also be used as an ingredient in the manufacture of cosmetics.

Organizations have been formed to aid in the distribution of organic honey to different distribution stores. The profits from the selling of organic honey tend to be high since the demand for organic honey is also high. This boosts the living standards of the farmers.

10.9 CHALLENGES FACING THE MARKETING OF ORGANIC LIVESTOCK AND THEIR PRODUCTS

One of the main challenges facing the marketing of organic livestock and their products is the issue of lack of awareness among the public. Most people still purchase conventional organic products and they do not believe in the fact that organic products are healthier than conventional livestock products. This challenge can be addressed through the establishment of forums that would educate the public on the importance of organic livestock and their products. The second challenge is the cost of practising organic livestock rearing. The availability of organic livestock feeds is still a challenge more so to the farmers who are in remote areas and who have limited pieces of land for them to be able to practice both livestock rearing and the production of Organic livestock feeds. This lowers the number of organic livestock products present in the market platforms.

The third challenge has been discussed severally in the above sections and it is the issue of substitution. Corrupt individuals substitute conventional livestock products for organic livestock products by packaging conventional livestock products in organic livestock products packages. This challenge can only be addressed by ensuring that the suppliers are certified to supply organic livestock products. The products should also be tested from time to time to ensure that they are purely organic.

More organizations should be set up to aid in the supply of organic livestock products. Technology should also be incorporated in the marketing of organic products through the invention of applications that specialize in the online marketing of organic livestock products. This avails organic livestock products to a wider market.

The other challenge facing the marketing of organic livestock products is the high prices of the products, which is contributed by the intensive labor required in organic farms. Government should support the innovation of mechanization techniques that would adhere to the rules of organic farming such as the maintenance of the environment and the Conservation of biodiversity. The mechanization techniques would aid in reducing the workforce required in the production of organic livestock and their produce. With this, the pricing of organic livestock products will be standardized and affordable to everyone.

10.10 CONCLUSION

Organic livestock production is the future of the world. It will ensure that human beings feed on quality and healthy animal products. Cases of ailments such as heart attacks, cancer among others will drop enabling the health personnel to focus on the treatment of other illnesses and save on the expensive cost incurred in the treatment of such illnesses (Vogl *et al.*, 2016). The production of organic livestock should be promoted all over the world, and organic livestock products made easily available to the public at a standardized price. Conventional livestock production ought to be slowly turned into organic livestock production. Markets to organic livestock products should be readily available.

References

1. Ankamah-Yeboah, I., Jacobsen, J. B., Olsen, S. B., Nielsen, M., & Nielsen, R. (2019). The impact of animal welfare and environmental information on the choice of organic fish: An empirical investigation of German trout consumers. *Marine Resource Economics*, *34*(3), 247-266.
2. Åkerfeldt, M. P., Gunnarsson, S., Bernes, G., & Blanco-Penedo, I. (2020). Health and welfare in organic livestock production systems—a systematic mapping of current knowledge. *Organic Agriculture*, 1-28.
3. Alonso, M. E., González-Montaña, J. R., & Lomillos, J. M. (2020). Consumers' concerns and perceptions of farm animal welfare. *Animals*, *10*(3), 385.
4. Bessei, W. (2018). Impact of animal welfare on worldwide poultry production. *World's Poultry Science Journal*, *74*(2), 211-224.
5. Buller, H., Blokhuis, H., Jensen, P., & Keeling, L. (2018). Towards farm animal welfare and sustainability. *Animals*, *8*(6), 81.
6. Broom, D. M., Galindo, F. A., & Murgueitio, E. (2013). Sustainable, efficient livestock production with high biodiversity and good welfare for animals. *Proceedings of the Royal Society B: Biological Sciences*, *280*(1771), 20132025.
7. Brito, L. F., Oliveira, H. R., McConn, B. R., Schinckel, A. P., Arrazola, A., Marchant-Forde, J. N., & Johnson, J. S. (2020). Large-scale

phenotyping of livestock welfare in commercial production systems: A new frontier in animal breeding. *Frontiers in Genetics, 11*.

8. Cambra-López, M., Aarnink, A. J., Zhao, Y., Calvet, S., & Torres, A. G. (2010). Airborne particulate matter from livestock production systems: A review of an air pollution problem. *Environmental pollution*, *158*(1), 1-17.
9. Castellini, C., Boggia, A., Cortina, C., Dal Bosco, A., Paolotti, L., Novelli, E., & Mugnai, C. (2012). A multicriteria approach for measuring the sustainability of different poultry production systems. *Journal of Cleaner Production*, *37*, 192-201.
10. Cornish, A., Raubenheimer, D., & McGreevy, P. (2016). What we know about the public's level of concern for farm animal welfare in food production in developed countries. *Animals*, *6*(11), 74.
11. Clonan, A., Wilson, P., Swift, J. A., Leibovici, D. G., & Holdsworth, M. (2015). Red and processed meat consumption and purchasing behaviours and attitudes: impacts for human health, animal welfare and environmental sustainability. *Public health nutrition*, *18*(13), 2446-2456.
12. Clark, B., Stewart, G. B., Panzone, L. A., Kyriazakis, I., & Frewer, L. J. (2016). A systematic review of public attitudes, perceptions and behaviours towards production diseases associated with farm animal welfare. *Journal of Agricultural and Environmental Ethics*, *29*(3), 455-478.
13. Chander, M., Bodapati, S., Mukherjee, R., & Kumar, S. (2011). Organic livestock production: an emerging opportunity with new challenges for producers in tropical countries. *Scientific and Technical Review of the Office International des Epizooties (Paris)*, *30*(3), 569-583.
14. Celi, P., Cowieson, A. J., Fru-Nji, F., Steinert, R. E., Kluenter, A. M., & Verlhac, V. (2017). Gastrointestinal functionality in animal nutrition and health: new opportunities for sustainable animal production. *Animal Feed Science and Technology*, *234*, 88-100.
15. D'Silva, J., & Webster, J. (Eds.). (2017). The meat crisis: Developing more sustainable and ethical production and consumption. Routledge,Pp-390.
16. Dippel, S., Leeb, C., Bochicchio, D., Bonde, M., Dietze, K., Gunnarsson, S., & Prunier, A. (2014). Health and welfare of organic pigs in Europe assessed with animal-based parameters. *Organic Agriculture*, *4*(2), 149-161.

17. de Jonge, J., & van Trijp, H. C. (2013). The impact of broiler production system practices on consumer perceptions of animal welfare. *Poultry science*, *92*(12), 3080-3095.
18. Donnik, I., Voronin, B., & Loretts, O. (2016). Production of organic agricultural products is an important area of "Green" economy. *Indian journal of science and technology*, *9*(14), 91512.
19. Durmic, Z., & Blache, D. (2012). Bioactive plants and plant products: Effects on animal function, health and welfare. *Animal feed science and technology*, *176*(1-4), 150-162.
20. Escribano, A. J. (2016). Organic livestock farming–challenges, perspectives, and strategies to increase its contribution to the agrifood system's sustainability–a review. *Organic farming–a promising way of food production'.(Ed. P Konvalina) pp*, 229-260.
21. Escribano, A. J., Ryschawy, J., & Whistance, L. K. (2019). Integrated crop–livestock systems with agroforestry to improve organic animal farming. In *Book chapter* (pp. 123-155). Burleigh Dodds Science Publishing.
22. Entz, M. H., & Thiessen Martens, J. R. (2009). Organic crop–livestock systems. *Organic farming: The ecological system*, *54*, 69-84.
23. Escribano, A. J. (2016). Organic livestock farming–challenges, perspectives, and strategies to increase its contribution to the agrifood system's sustainability–a review. *Organic farming–a promising way of food production'. (Ed. P Konvalina) pp*, 229-260.
24. Gunnarsson, S., & Mie, A. (2018). 12. Organic animal production–a tool for reducing antibiotic resistance?. *Wageningen Academic P ublishers*, 93.
25. Grunert, K. G., Sonntag, W. I., Glanz-Chanos, V., & Forum, S. (2018). Consumer interest in environmental impact, safety, health and animal welfare aspects of modern pig production: Results of a cross-national choice experiment. *Meat science*, *137*, 123-129.
26. Honkanen, P., & Olsen, S. O. (2009). Environmental and animal welfare issues in food choice. *British Food Journal*, *111(2-3)*, 293-309.
27. Ivemeyer, S., Smolders, G., Brinkmann, J., Gratzer, E., Hansen, B., Henriksen, B. I., & Walkenhorst, M. (2012). Impact of animal health and welfare planning on medicine use, herd health and production in European organic dairy farms. *Livestock Science*, *145*(1-3), 63-72.

28. Jespersen, L. M., Baggesen, D. L., Fog, E., Halsnæs, K., Hermansen, J. E., Andreasen, L.,& Halberg, N. (2017). Contribution of organic farming to public goods in Denmark. *Organic Agriculture*, *7*(3), 243-266.
29. Kiefer, L., Menzel, F., Bahrs, E. (2014). The effect of feed demand on greenhouse gas emissions and farm profitability for organic and conventional dairy farms. *Journal of Dairy Science, 97*(12), 7564-7574.
30. Kleter, G., McFarland, S., Bach, A., Bernabucci, U., Bikker, P., Busani, L., & Einspanier, R. (2018). Surveying selected European feed and livestock production chains for features enabling the case-specific post-market monitoring of livestock for intake and potential health impacts of animal feeds derived from genetically modified crops. *Food and Chemical Toxicology*, *117*, 66-78.
31. Kongsted, H., & Sørensen, J. T. (2017). Lesions found at routine meat inspection on finishing pigs are associated with production system. *The Veterinary Journal*, *223*, 21-26.
32. Krieger, M., Sjöström, K., Blanco-Penedo, I., Madouasse, A., Duval, J. E., Bareille, N.,& Emanuelson, U. (2017). Prevalence of production disease related indicators in organic dairy herds in four European countries. *Livestock Science*, *198*, 104-108.
33. Leeb, C., Hegelund, L., Edwards, S., Mejer, H., Roepstorff, A., Rousing, T.,& Bonde, M. (2014). Animal health, welfare and production problems in organic weaner pigs. *Organic agriculture*, *4*(2), 123-133.
34. Lu, C. D., Gangyi, X., & Kawas, J. R. (2010). Organic goat production, processing and marketing: Opportunities, challenges and outlook. *Small Ruminant Research*, *89*(2-3), 102-109.
35. Lovarelli, D., Bacenetti, J., & Guarino, M. (2020). A review on dairy cattle farming: Is precision livestock farming the compromise for an environmental, economic and social sustainable production?. *Journal of Cleaner Production*, *262*, 121409.
36. Manuelian, C. L., Penasa, M., da Costa, L., Burbi, S., Righi, F., & De Marchi, M. (2020). Organic Livestock Production: A Bibliometric Review. *Animals*, *10*(4), 618.
37. Mench, J. A. (2008). Farm animal welfare in the USA: Farming practices, research, education, regulation, and assurance programs. *Applied Animal Behaviour Science*, *113*(4), 298-312.

38. Mingmongkolchai, S., & Panbangred, W. (2018). Bacillus probiotics: an alternative to antibiotics for livestock production. *Journal of applied microbiology*, *124*(6), 1334-1346.
39. Mie, A., Andersen, H. R., Gunnarsson, S., Kahl, J., Kesse-Guyot, E., Rembiałkowska, E., & Grandjean, P. (2017). Human health implications of organic food and organic agriculture: a comprehensive review. *Environmental Health*, *16*(1), 1-22.
40. Miranda-De La Lama, G. C., Estévez-Moreno, L. X., Sepulveda, W. S., Estrada-Chavero, M. C., Rayas-Amor, A. A., Villarroel, M., & María, G. A. (2017). Mexican consumers' perceptions and attitudes towards farm animal welfare and willingness to pay for welfare friendly meat products. *Meat science*, *125*, 106-113.
41. Nalubwama, S. M., Mugisha, A., & Vaarst, M. (2011). Organic livestock production in Uganda: potentials, challen Neethirajan, S., Tuteja, S. K., Huang, S. T., & Kelton, D. (2017). Recent advancement in biosensors technology for animal and livestock health management. *Biosensors and Bioelectronics*, *98*, 398-407.ges and prospects. *Tropical Animal Health and Production*, *43*(4), 749-757.
42. Padel, S. (2019). The principles of organic livestock farming. In *Improving Organic Animal Farming* (pp. 13-31). Burleich Dodds Science Publication.
43. Papatsiros, V. G. (2011). Impact of animal health management on organic pig farming in Greece. *Biotechnology in Animal Husbandry*, *27*(1), 115-125.
44. Pérez, I. A. V., Toral, J. N., Vázquez, Á. T. P., Hernández, F. G., Ferrer, G. J., & Cano, D. G. (2019). Potential for organic conversion and energy efficiency of conventional livestock production in a humid tropical region of Mexico. *Journal of Cleaner Production*, *241*, 118354.
45. Prunier, A., Lubac, S., Mejer, H., Roepstorff, A., & Edwards, S. (2014). Health, welfare and production problems in organic suckling piglets. *Organic agriculture*, *4*(2), 107-121.
46. Röös, E., Mie, A., Wivstad, M., Salomon, E., Johansson, B., Gunnarsson, S., & Watson, C. A. (2018). Risks and opportunities of increasing yields in organic farming. A review. *Agronomy for Sustainable Development*, *38*(2), 1-21.
47. Sorge, U. S., Yamashita, S., & Pieper, L. (2019). Bovine veterinarians' perspective on organic livestock production in the USA. *Veterinary Record*, *184*(12), 384-384.

48. Sonoda, Y., Oishi, K., Chomei, Y., & Hirooka, H. (2018). How do human values influence the beef preferences of consumer segments regarding animal welfare and environmentally friendly production?. *Meat science*, *146*, 75-86.
49. Sossidou, E. N., Dal Bosco, A., Castellini, C., & Grashorn, M. A. (2015). Effects of pasture management on poultry welfare and meat quality in organic poultry production systems. *World's poultry science journal*, *71*(2), 375-384.
50. Souillard, R., Répérant, J. M., Experton, C., Huneau-Salaun, A., Coton, J., Balaine, L., & Le Bouquin, S. (2019). Husbandry practices, health, and welfare status of organic broilers in France. *Animals*, *9*(3), 97.
51. Sorge, U. S., Yamashita, S., & Pieper, L. (2019). Bovine veterinarians' perspective on organic livestock production in the USA. *Veterinary Record*, *184*(12), 384-384.
52. Sutherland, M. A., Webster, J., & Sutherland, I. (2013). Animal health and welfare issues facing organic production systems. *Animals*, *3*(4), 1021-1035.
53. Sundrum, A. (2012). 6 Health and Welfare of Organic Livestock and Its Challenges. *Organic Meat Production and Processing*, *53*.
54. Smit, L. A., & Heederik, D. (2017). Impacts of intensive livestock production on human health in densely populated regions. *GeoHealth*, *1*(7), 272-277.
55. Tallentire, C. W., Edwards, S. A., Van Limbergen, T., & Kyriazakis, I. (2019). The challenge of incorporating animal welfare in a social life cycle assessment model of European chicken production. *The International Journal of Life Cycle Assessment*, *24*(6), 1093-1104.
56. Thornton, P. K. (2010). Livestock production: recent trends, future prospects. *Philosophical Transactions of the Royal Society B: Biological Sciences*, *365*(1554), 2853-2867.
57. Vaarst, M., & Alrøe, H. F. (2012). Concepts of animal health and welfare in organic livestock systems. *Journal of agricultural and environmental ethics*, *25*(3), 333-347.
58. Vanhonacker, F., Van Poucke, E., Tuyttens, F., & Verbeke, W. (2010). Citizens' views on farm animal welfare and related information provision: Exploratory insights from Flanders, Belgium. *Journal of Agricultural and Environmental Ethics*, *23*(6), 551-569.

59. Von Borell, E., & Sørensen, J. T. (2004). Organic livestock production in Europe: aims, rules and trends with special emphasis on animal health and welfare. *Livestock Production Science*, *90*(1), 3-9.

60. Vogl, C. R., Vogl-Lukasser, B., & Walkenhorst, M. (2016). Local knowledge held by farmers in Eastern Tyrol (Austria) about the use of plants to maintain and improve animal health and welfare. *Journal of Ethnobiology and Ethnomedicine*, *12*(1), 1-17.

61. Wilhelmsson, S., Yngvesson, J., Jönsson, L., Gunnarsson, S., & Wallenbeck, A. (2019). Welfare Quality® assessment of a fast-growing and a slower-growing broiler hybrid reared until 10 weeks and fed a low-protein, high-protein or mussel-meal diet. *Livestock Science*, *219*, 71-79.

INDEX

L

M

N

O

P

Printed in the United States
by Baker & Taylor Publisher Services